Green Energy and Technology

T0138068

For further volumes:
http://www.springer.com/series/8059

Srinivasan Sunderasan

Rational Exuberance
for Renewable Energy

An Economic Analysis

 Springer

Dr. Srinivasan Sunderasan
Verdurous Solutions Private Limited
145, Navillu Road 7th Cross Kuvempunagar
570023 Mysore, Karnataka
India
e-mail: sunderasan@yahoo.com

ISSN 1865-3529   e-ISSN 1865-3537

ISBN 978-1-4471-2630-0   ISBN 978-0-85729-212-4 (eBook)

DOI 10.1007/978-0-85729-212-4

Springer London Dordrecht Heidelberg New York

British Library Cataloging in Publication Data
A Catalogue record for this book is available from the British Library

Cover design: eStudio Calamar, Berlin/Figueres

Printed on acid-free paper

Springer is part of Springer Science+Business Media (www.springer.com)

Foreword

I know Dr. Sunderasan and his work from his days as a Ph.D. student in Vienna. And from the very beginning two things characterize him, one personal (very eager to learn and to engage in research) and the second concerns his fresh views on alternative energy in particular solar energy (PV, primarily) that contrast so starkly the thinking in the industrialized world. His propositions are: (i) alternative energy and solar power in particular must succeed in the market (and not in the game for subsidies) and (ii) that developing countries with their lack or unreliable supply due to their poor infrastructure, the lack of a grid, the many blackouts, the costly technologies diesels are the area where solar power should have a comparative advantage. This contrasts the fact that Germany is proud as the leading in number of solar panel installations yet having world class infrastructure and a summer a German poet of the 19th century called 'mild winters' (although global warming has changed that a bit, but just writing this foreword at the beginning of September, we had snow fall down to 1000m elevation). Of course, this success is financed by hefty subsidies via (much acclaimed) high feed in tariffs generating a money pump for its investors. It is obvious that an efficient deployment of solar energy must start as Dr. Sunderasan suggests in his works contrasting a lot of conventional wisdom.

Univ. Prof. Dr. Franz Wirl
Chair of Industry, Energy & Environment
Faculty of Business, Economics & Statistics
University of Vienna
Vienna, Austria

Preface

Growing environmental consciousness, widely reported 'doomsday' predictions by acknowledged experts, geopolitical concerns relating to uninterrupted supplies of fossil fuels and uncertainties associated with fuel prices, stimulus packages and incentive schemes announced by several governments have all led to the rapid and strong growth of the Renewable Energy (RE) industry. Policy makers expect to trigger in its wake, a virtuous cycle of enlarged production, scale economies, lower prices, and even more rapid market growth. Some forms of renewable energy are now grid-competitive in certain market conditions. While experts believe that several new and 'natural' markets could sustain even higher rates of renewable energy penetration, growth has seldom lived up to expectations, targets have frequently been missed and numerous market related barriers remain.

Yet there is little economic analysis of renewable energy relative to other industries of similar scope and stature. The main renewable energy resources face a problem because of their intermittency (the wind does not always blow nor does the sun shine all the time) and this has not always been factored into estimation of their potential. Few researchers have actually applied the fundamentals of economic theory to furthering the cause of Renewable Energy (interchangeably used with such other terms as "sustainable energy", "clean energy" etc.) with a view to helping it grow into a self-sustaining industry sector. Several policy makers and consultants, for instance, have implicitly assumed in the past that consumers would pay higher prices for energy from solar PV or wind farms almost exclusively based on their stated preferences. Likewise, biofuels viz., ethanol and bio-diesel are assumed to gross higher unit prices than their petroleum derived counterparts exclusively basing on their superior environmental credentials. Over time, the dependence on grants and subsidies has only been enhanced and this supposedly 'sunrise' industry has remained in the shadows.

This book is an attempt to blend classical economic theory with the realities of the RE industry and to identify incentive structures contributing to the success or otherwise of project implementation involving renewable sources and appropriate technologies. It is an attempt to shoot straight at the underlying issues that encourage or plague widespread dissemination of RE technologies. The chapters carry meticulously researched articles published by reputed international peer-reviewed journals, with their respective list of references, foot notes and end notes pointing towards sources of additional data, offering clarifications and illustrations or listing exceptions. In many instances, data from India, a setting which provides an excellent backdrop to study various issues involved, are used to illustrate and to drive the arguments.

Karnataka, August 2010   Dr. Srinivasan Sunderasan

Table of Contents

Terms and Abbreviations

Alternative energy	Often used interchangeably with 'Renewable Energy', energy derived from sources that do not consume natural assets and those that do not harm the environment
Bagasse	Moisturized fiber residue from crushing cane; traditionally, a sugar mill waste
Bio-fuel	Fuel derived from plants and organic sources as opposed to petro-diesel and gasoline, which are petroleum distillates
Bureaucratic entrenchment	The bureaucracy makes its own removal costly by hoarding information, front-loading the benefits of a policy etc.
CER	Certified emission reduction ("carbon credit")
CHP	Combined heat and power ("cogeneration" and sugarcane-cogeneration when applied to a sugar mill)
Commoditized product	A generic product with vendors competing almost exclusively on price
Decentralized generation	Power generated at or close to the point of consumption; surplus power is frequently fed to the utility grid and is paid for at a rate referred to as the feed-in-tariff
Development	The enlargement of people's choices, as defined by the United Nations Development Program (UNDP), Human Development Report (HDR)
Economic profit	The rate of return in excess of the return earned on the nearest feasible alternative application of the investor's capital.
Economic rent	The additional amount paid to a factor of production over and above its market-determined

	price, owing to its market power derived from its scarcity
Electrified home	A dwelling unit which has access to the utility grid
Energy plantation	Farming of 'energy crops' such as corn, jatropha etc, that could potentially yield bio-fuels
Exogenously determined	Prices and other parameters determined externally, for instance, by government notification: opposite of market determined
Fiscal incentive	Tax credits, accelerated depreciation and other benefits offered by governments to encourage growth of specified industry sectors
FiT	Feed-in Tariff: the price per unit of electricity the utility is obligated to pay to private generators; typically used in the context of power generated from roof-top solar photovoltaic systems
GEF	Global Environmental Facility (www.thegef.org)
Grid-parity	Alternative technologies generating power at the same costs as traditional, centralized power plants
Grid-supplied power	Power generated by a centralized power plant viz., thermal, hydro, nuclear etc., and supplied to consumers by the utility grid network
IFC	International Finance Corporation, member of the World Bank Group (www.ifc.org)
Incentive (structure)	Factor (or set of factors) that enables or motivates a particular course of action
Inclusive development	Ensuring equality of opportunity for all classes of society
Infant industry	New industries that often seek protection from international trade and foreign competition
INR	Indian Rupee
IREDA	Indian Renewable Energy Development Agency (www.ireda.in)
LCA	Life Cycle Assessment
LED	Light Emitting Diode
LPD	Liter per day
Mature industry	A sector where the product and its application is clearly defined and understood and where process improvements dominate product improvements
MFI	Microfinance Institution
Micro-credit	Small-sized, often unsecured loans
Monopoly rent	Higher profit margins cornered by a firm by virtue of being a sole supplier of a particular good or service

Monopsony	A market dominated by a single buyer, as with electricity utilities buying power from independent power producers
Razor-and-blade pricing	Or complementary-product-pricing: a strategy involving distribution of the base hardware at break-even or even a deeply discounted price, expecting to earn positive economic returns on add-on components and services required to drive the hardware
Renewable Energy Source	Natural resource which when employed is not necessarily depleted viz., driving multiple turbines with the same stream of water (i.e. 'non-rival' consumption by any single generator)
RERED	Renewable Energy for Rural Economic Development: a World Bank supported program in Sri Lanka
R&D	Research and Development
RET	Renewable Energy Technology: technology that harnesses naturally occurring, theoretically inexhaustible sources of energy, as biomass, solar, wind, tidal, wave, and hydroelectric power, that are not derived from fossil or nuclear fuels
Revealed preferences	Consumption decisions and actual payments made by consumers: as opposed to stated preferences which are statements of intention
SAP	State advised price to be paid to farmers supplying sugarcane to a sugar mill
Self-liquidating asset	An investment that generates cash-flows sufficient to offset the upfront costs
SHS	Solar Home System: domestic system which provides lighting, and supports other appliances like fans, radio, mobile phone charging etc.
Spillover (also externality)	An economic side-effect, often an unintended consequence
Stated preferences	Consumption decisions, willingness to pay etc., as voiced by potential consumers; statements of intent
Subsidy(ies)	Money paid, usually by governments, to keep prices below free-market levels with a view to keeping weak industry sectors from going bust
Sugar year	October – September in India
Sunrise industry	New, emerging growth industry that is expected to be a strong sector in the future

Sustainable development	Making more choices available at present, while ensuring that doing so does not curtail choices that could be made available to future generations
SWH	Solar Water Heating System
TCD	Metric tons of sugarcane crushed per day, in a sugar mill
Technology / sector agnostic	Policy or preferences that are not sector or technology specific
Zero economic profits	A rate of return that is no greater than the 'normal' return earned on the nearest feasible alternative application of the investor's capital

Chapter 1
Infant Industry and Incentive Structures

> *Sir Humphrey: "Bernard, subsidy is for art…for culture. It is not to be given to what the people want, it is for what the people don't want but ought to have."*
> Yes Minister Television series
> "The Middle-Class Rip-Off", 23 December 1982

The *infant industry* argument has persisted through most of industrial era history in one form or the other. While in the mercantilist era, government intervention through subsidies, tariffs, quotas and other fiscal measures, and through non-tariff barriers to imports, has been justified by the face-off between the "developed" and the "developing" worlds, the virtual collapse of national borders and the consequent vertical-disintegration-of-production-processes has spawned new theories. The lines of reasoning have shifted to "mature industries versus emerging technology" or between the so called *entrenched* "old economy" sectors against the *sunrise* "new economy" sectors. On grounds that the mature industry incumbent is yet being propped up as with the fossil fuel industry or the power sector, or that the incumbent has been supported in the past, as with the fixed-line telecom sector, intervention and protection are slated to "level the playing field" for the nascent challenger. In several industry sectors, traditional incentives for import substitution and such inward-looking policies have been replaced by policies encouraging export and international competition, though the optimal degree of intervention by the state is still being debated [5]. Political-economy arguments often go beyond the immediate industry-specific goals to include early-mover advantages, job creation, and self-sufficiency in times of geo-political uncertainty and the like. It is, now, generally accepted that governments need to *temporarily* support and protect industry sectors which show *promise* until they grow strong *enough* to compete on their own strengths.

Children wrapped in cotton wool are seldom proof from contamination. The tenure of infancy of individual sectors is uncertain and subjective. 'Promise' of

S. Sunderasan, *Rational Exuberance for Renewable Energy*,
Green Energy and Technology, DOI: 10.1007/978-0-85729-212-4_1,
© Springer-Verlag London Limited 2011

the future is always subject to numerous assumptions relating to future states of the world and seldom pans out as originally envisaged. While Korea and select East Asian countries are said to stand out, [9] there have been numerous cases of inefficient firms or entire industry sectors sustained by subsidies, protectionist measures and such other assistance from governments, only to fall by the wayside the moment the markets are opened up to competition. Das and Srinivasan [4] study the duration of firms in an infant industry and observe that larger sizes at entry lead to higher probabilities of exit from infancy and longevity thereafter, and that diversification and public ownership do not necessarily influence survival rates.

Further, in the globalized world of finance, business propositions that hold promise should find it relatively straightforward to mobilize equity and debt capital, almost independent of geographic location. Bell et al. [2] conclude that in less developed countries, infant firms have experienced slow productivity growth and more frequently than not, have failed to achieve international competitiveness. The authors also report that conscious efforts need to be invested in acquiring capabilities for continuous technological change: a rather paradoxical obligation, in reality, since such technological upgrade and evolution would mean that on the one hand, firms forego current profits, and on the other, subsequent to such upgrades the firms in question outgrow the cozy comfort of the protectionist blanket. In fact, Saure [15] concludes that the infant industry argument is flawed and that 'protected' domestic producers are likely to substitute advanced technologies with low-growth 'traditional technology', thereby ensuring continuity but inhibiting learning and advancement.

Panda and Ramanathan [14] propose a methodology to assess the technological capabilities of a firm:

1. Identification of value-addition stages performed by a firm;
2. Determination of technological capabilities needed at these stages;
3. Development of indicators for assessing the identified technological capabilities;
4. Benchmarking the capabilities assessed with a state-of-the-art firm; and
5. Analysis of the reasons for the technological capability gap between the firm being studied and the state-of-the-art firm.

In analyzing when and how infant industry should be protected, Melitz [11], demonstrates that the decision to protect a particular industry should depend on the "industry's leaning potential, the shape of the learning curve, and the degree of substitutability between domestic and foreign goods". Ohyama et al. [12] go on to argue that "the learning rate depends on the quality of ideas, not the scale of the industry" and that a competitive environment fosters innovation-led growth. For instance, the internationally competitive Japanese textile industry was set up without government protection, after attempts at subsidizing firms failed rather comprehensively. Worse, transactional and short-term protectionist measures tend to be ineffective, and aggravate rather than ameliorate, long-term structural inadequacies. Grabowski [8] believes that the effectiveness of government policy aimed at supporting industrialization depends on the ability of the state to discipline firms, the credibility of such policy and crucially, the size of the domestic markets. In summary, decisions relating to protection, import substitution and technology acquisition, should be influenced by "overall development strategy, capabilities in terms of skills and industrial infrastructure, the size of the home market and appreciation of characteristics of the international capital goods market" [3].

1.1 Incentive Structures

Economics is about visualizing basic, universally applicable principles, patterns and sequences of events underlying apparently complex developments. It is now widely acknowledged that a well regulated competitive market environment makes efficient use of resources and encourages rapid innovation in contrast to state controlled production. Intensive intervention in production and pricing decisions distorts consumption patterns, while contributing to deterioration in product and service quality and stagnation among the work force. Worse, it also leads to inconsistencies in the treatment meted out to various players and to bureaucratic entrenchment while creating perverse incentives culminating in sustaining inefficient producers.

Intuitively, it is obvious that skill development among workers in inward-looking manufacturing is inhibited by the reduced exposure and fewer interactions outside of the immediate work sphere. Technological innovation is a systemic phenomenon and obviously cannot be brought about by diktat. Liu and White [10] have developed a generic framework encompassing five fundamental activities—Research and Development (R&D), implementation, end-use, education, linkage—to analyze the system-level characteristics in introducing, diffusing and exploiting technological innovation. Okada [13] asserts that the new institutional mechanisms in a liberalized environment play a "crucial role" in upgrading the "supply chain into a learning chain".

Locating similar business operations in clusters also leads to learning by 'spillover' of knowledge and practices in informal interactions among workers from different units. Altenburg and Meyer-Stamer [1] categorize industrial clusters into three groups, each with reasons for coming into existence, their respective time profiles and embedded challenges.

"Survival Clusters" of micro and small-scale enterprises are not inherently competitive owing to acute limitations in entrepreneurial competence and dynamism, but are important sources of employment, especially in labor-abundant third-world countries. It should be possible to identify the potential for consolidation among the enterprises, and to chart out skill upgrades and capital investments required to help enhance the employment opportunity in such clusters.

Mass producers, often with little competition and indifferent product attributes have survived and prospered within closed economies. Such clusters need to be encouraged to invest in learning, product upgrades and continuous innovation if they are to survive in competitive settings.

Transnational clusters are organized around home-country concepts viz., the Korean *Chaebol* or the Japanese *Kiretsu*, bringing with them the best practices in manufacturing and supply chain management. Domestic firms could benefit immensely from associating with such corporations and clusters and upgrade their management processes, skill levels and product attributes.

Designing the right incentive structures for individuals and firms is far from straightforward. Pointing out that the first aircraft built by the Wright Brothers was far cheaper to build compared to the costs of even marginally upgrading a modern day jetliner, William Baumol suggests that breakthroughs are "breathtakingly simple", and often less complicated though they are far more sporadic and uncertain compared to routine innovations. Businesses have tried to regularize such inventiveness through routine investments in research and development, with a view to earning stable and predictable returns. The contrast in incentives is stark: for an entrepreneur investing his labor of love, pursuing a pet idea over a lifetime, a "touch of madness" is probably an essential qualification; on the other hand, big firms seeking to commercialize ideas employ professionals who look out for competitive wages, [6].

Traditionally, high-income economies have invented new product while developing countries have manufactured and traded goods in large measure. Over time, such assimilation of know-how and continued improvement of both product and process has blurred the distinction between the 'home of invention' and the 'base for production'. Creativity knows no boundaries: creating appropriate incentive structures is far more challenging in the new world. Industrial policy as practised in the decades past has been generally discredited because governments have backed select industries and firms, almost at random. While there have been a few noteworthy success stories, more often, scarce resources have been squandered. Experts are coming around to the view that in the new world order, industrial policy should be "agnostic, promoting industries or techniques the country has never tried before", [7] as opposed to backing existing technologies or players. In providing encouragement for industrial development, and in incubating business ventures, governments are advised to "spread bets widely, monitor results rigorously and cut losses ruthlessly". Following the venture capital model of rationing scarce resources, governments need to design industrial policy that would be quick to shed losers, even if the technocrats involved are unable to pick winners!

References

1. Altenburg T, Meyer-Stamer J (1999) How to promote clusters: policy experiences from Latin America. World Dev 27(9):1693–1713
2. Bell M, Ross-Larson B, Westphal LE (1984) Assessing the performance of infant industries. J Dev Econ 16(1–2):101–128
3. Chesshire J, Surrey J (1985) Energy technology acquisition for third world development. Energy Policy 13(4):316–319

4. Das S, Srinivasan K (1997) Duration of firms in an infant industry: the case of Indian computer hardware. J Dev Econ 53(1):157–167
5. Dervis K, Page M Jr (1984) Industrial policy in developing countries. J Comp Econ 8(4): 436–451
6. The Economist (2006) Economic focus: searching for the invisible man. The Economist Print Edition, 9th March
7. The Economist (2009) Industrial design. The Economist Print Edition, 1st October
8. Grabowski R (1994) The successful developmental state: where does it come from? World Dev 22(3):413–422
9. Lee J (1997) The maturation and growth of infant industries: the case of Korea. World Dev 25(8):1271–1281
10. Liu X, White S (2001) Comparing innovation systems: a framework and application to China's transitional context. Res Policy 30(7):1091–1114
11. Melitz MJ (2005) When and how should infant industries be protected? J Int Econ 66(1): 177–196
12. Ohyama A, Braguinsky S, Murphy KM (2004) Entrepreneurial ability and market selection in an infant industry: evidence from the Japanese cotton spinning industry. Rev Econ Dyn 7(2):354–381
13. Okada A (2004) Skills development and interfirm learning linkages under globalization: lessons from the Indian automobile industry. World Dev 32(7):1265–1288
14. Panda H, Ramanathan K (1996) Technological capability assessment of a firm in the electricity sector. Technovation 16(10):561–588
15. Saure P (2007) Revisiting the infant industry argument. J Dev Econ 84(1):104–117

Chapter 2
Renewable Energy Technology: Market Development, Subsidy Policy and the Enlargement of Choice

Begin with the end in mind.
Stephen R. Covey
The Seven Habits of Highly Effective People, 1989

The introductory chapter lays out the theme, and more importantly, sets the tone for the ensuing portions of this book. In keeping with traditional wisdom the next chapter stresses that 'direction is more important than distance covered' and is intended to exhort policy makers to 'begin with the end in mind' so as to define and achieve the goals of positive development policy within prescribed time limits and within budgeted costs. Given the different approaches to market stimulation adopted for each, the Indian markets for solar photovoltaic (PV) and solar thermal (solar water heating systems: SWH) systems are compared. The conclusions relating to the delivery channels for product, service and financing, the incentive structures and most significantly, the overarching targets relating to consumer empowerment would, however, apply across technology options, market geographies and consumer segments.

Development is the enlargement of people's choices.[1] Government policy that restricts choice is, by extension, regressive. Governments might choose to subsidize certain sections of consumers for political reasons or as a part of development strategy. In most developing countries, politics often affect the continuance,

This chapter is largely based on Srinivasan, Sunderasan (2009) Subsidy policy and the enlargement of choice. Renewable and Sustainable Energy Reviews 13:2728–2733.

[1] The First UNDP—Human Development Report, 1990, http://hdr.undp.org/en/reports/global/hdr1990/.

S. Sunderasan, *Rational Exuberance for Renewable Energy*,
Green Energy and Technology, DOI: 10.1007/978-0-85729-212-4_2,
© Springer-Verlag London Limited 2011

distribution, geographic expansion and eventual termination, if at all, of subsidies [7]. Although discounted prices increase consumer demand for normal goods, they could lead to supply-side distortions, consequently restricting people's choices. A commodity or a service is deemed expensive simply because of consumers' low willingness to pay relative to substitutes, and subsidies could potentially end up compelling them to choose the less preferred alternative while simultaneously dissuading suppliers from serving such markets.

Even when launched with the most noble of intentions, inappropriately designed or untargeted subsidy programs could, thus, end up being counterproductive. Expanding and improving markets, therefore, involves not just assessing the potential for a technology package to contribute towards development goals, but also building efficient and locally appropriate service delivery channels and most importantly, accounting for political constraints [20]. The policy instruments employed are shaped by the political process that Jacobson and Lauber [11] refer to as the 'battle over institutions' involving among others, the parliament, advocacy coalitions, local governments, competing commercial interests and the like.

A product or service is considered expensive or otherwise when compared to its next best alternative. Power from central generating stations supplied through utility grids is subsidized, especially for residential users, in many developing countries. The penetration of renewable energy technologies (RET) in most developing countries is impeded by sub-optimal pricing for power, continuing cross-subsidies [6] and retarded power sector reform, attributed mainly to the instability of policy-makers, poor overall acceptance of the reforms, slow adaptation and poor transition management [3]. The savings accruing from investments into energy efficient gadgets or into decentralized RET are minimal in such situations. Rationalizing power tariffs is projected to encourage energy conservation behavior and investments in energy efficiency while favoring distributed generation [18]. Others argue that in addition to streamlining the power sector, a national renewable energy policy is a vital prerequisite to translate customer choice into a larger market share for non-conventional energy technologies such as solar photovoltaic (PV) and thermal applications [19].

Large volume offtake of RE Technologies such as solar Photovoltaics has been brought about by favorable tariff regimes as in Japan and Germany. Yet the market collapses as soon as the incentives are pared down, as in Spain in 2008–2009. Wind energy is considered competitive with centrally generated power but suffers from intermittency and the need for reliable transmission infrastructure to convey the power from the generation nodes to load centers. In addition to favorable policy, renewable energy technologies (RET), like other technological advances require a period of nurturing diffusion and scaling-up to attain the price/performance ratios required to appeal to larger segments of the market. On the demand side, clean energy markets progress through 'education', 'policy support', 'standards', 'demonstration' and 'industry involvement' stages [13] to eventually morph into demand-driven markets.

2.1 Donor Experiences with Market Making Programs

- Grameen Shakti, the energy service company floated by Grameen Bank of Bangladesh has exploited the accumulated business acumen, the branch network and existing client base to create a successful solar energy (solar PV home lighting systems) business. Grameen Shakti has accessed refinance and grants from the World Bank's Rural Electrification and Renewable Energy Development Project through the Infrastructure Development Company.[2] The grants help bring down the first costs while the refinance provides requisite liquidity to expand the availability of micro-credit for the solar energy systems.
- Rated "satisfactory", the highest rank awarded, by the Independent Evaluation Group of the World Bank [23], the ongoing Renewable Energy for Rural Economic Development in Sri Lanka,[3] provides refinance and a co-financing grants to privately owned financial intermediaries, who in turn, extend retail credit for the procurement of solar energy systems and other technology packages. The program prescribes technical specifications for components such as photovoltaic modules, batteries, etc., while vendors are at liberty to configure systems to cater to market demand.
- Alternative schemes to encourage intermediaries to extend credit in rural areas include grants and default guarantees, usually extended by third parties or donor agencies as those provided by the Global Environment Facility (GEF)/International Finance Corporation (IFC) under the Photovoltaic Market Transformation Initiative (PVMTI). The default offset is viewed as a risk mitigation measure, and thus helps expand the availability of credit for the procurement of solar energy systems.
- The government of India provides subsidies to enhance the competitiveness of solar PV and solar thermal systems, which have, over the years, conditioned the supply-side of the industry.[4] For instance, the Indian Renewable Energy Development Agency (IREDA) has employed the leasing model, wherein financial intermediaries commonly referred to as non-banking finance companies, are provided with long-term, low-cost refinance for leased solar energy equipment.

2.2 Attempts at Market Making: An Illustration

Experience from such programs leads one to conclude that to achieve a measurable impact on market growth and to reach a diverse cross-section of users, retail financing terms need to be flexible. It is often mentioned that for the median

---

[2] www.idcol.org.

[3] www.energyservices.lk.

[4] India is not isolated in this regard. Spurred by low power prices, South Africa's aggregate demand for energy is slated to double in a decade. Observing that close to 20–30% of electricity use is dedicated to heating water, Eskom, the national utility has initiated "aggressive promotion of solar water heating systems" through a Rand 2 billion subsidy program [17].

consumer of decentralized energy services, (Rupees) 10 a day is more affordable than (Rupees) 300 a month, even as most policy makers, consultants and traditional bankers might fail to appreciate the distinction. Private sector financial intermediaries functioning within an unbiased competitive environment are most suited to offering innovative product-credit packages. In general, subsidies should be phased out simultaneously with strengthening of the institutions involved which implies *inter alia* that strengthening the institutions involved should be a primary objective. The framework of incentives for the participants should be developed with a view to ensuring sustainability of the value chain so as to ensure prompt post-installation service and help service providers benefit from resulting network effects.

A characteristic feature of the electricity sector in India is the unreliability [14] and power cuts are an acknowledged "way of life", with the demand–supply gap progressively widening [8]. Consequently, unlike in other developing countries where solar PV systems generally make their way into unelectrified homes, in India, such systems are acquired in large measure by households *with* access to the grid. Here again, electrification is generally measured by number of households having access to the grid, irrespective of the actual supply of electric power to these households. This distinction is more than just statistical: in rural areas and tail-ends of utility grids stand-alone as well as grid-tied renewable energy technologies compete against more expensive alternatives viz., dry cells, candles, etc., as opposed to grid-supplied electric power (whose supply is often unreliable), and hence could be rendered viable with appropriately structured financing.

Similarly, the solar water heaters (SWH) are employed to reduce dependence on grid supplied electricity and hence, there is a substantial overlap in the target markets for the solar PV lighting systems and solar water heaters in smaller towns and peri-urban areas. To illustrate the differences in approaches adopted to promote renewable energy technologies, the capital subsidy provided to encourage the deployment of solar PV systems ('solar home systems: SHS'), and the interest subsidy provided to support the dissemination of solar water heating systems (SWH) are compared. This is followed by an analysis of the channels used to route such subsidies and of the reasons for the outstanding and sustained success of one scheme relative to the other.

2.2.1 Solar Photovoltaic Home Lighting Systems

Solar photovoltaic technology is considered to be ideally suited to illuminate and to pump water at remote locations and dispersed settlements, where extension of the utility grid or making other alternatives available could be technically infeasible or economically unviable [4]. However, low customer density in a given service territory makes sales, installation, service and payment collection expensive and difficult, giving rise to transaction costs which are in the order of 30% of the total system costs [24]. This reduces affordability, undermines sustainability of

Table 2.1 Year wise targets and installations of SPV systems

Year	2002–2003		2003–2004		2004–2005		2005–2006		2006–2007	
INR (crore)	53.00	41.24	37.00	28.65	23.00	12.12	25.00	23.55	29.50	47.66
SHS (no.)	50,000	28,430	50,000	11,870	0	34,844	42,000	9,727	60,000	23,033

systems and diminishes the effect of the progressive reduction in PV system prices. Further, provision of credit in remote and rural areas is often perceived to be expensive, even unviable, owing to high collection costs and on account of limited collateral security that can be offered in the context. Service provision is challenging, especially on islands and remote areas with restricted access.

In contrast, the Grameen experience and the resulting micro-credit revolution have illustrated that availability of credit provides a sustained improvement in the quality of life. Ostensibly, creation of viable and sustainable sources of consumer finance is vital for the sustainability of renewable energy projects [9]. The Indian SPV Demonstration and Utilization Program is implemented through the state nodal agencies of the federal Ministry of New and Renewable Energy[5] and through select non-governmental organizations. The program has been operational since 1993–1994 and serves to provide subsidies to facilitate the purchase/installation of solar home systems, street lighting systems, and similar SPV applications.

The Ministry has laid down detailed specifications of all SPV lighting and other systems and prototypes are tested and approved by authorized test centers. The five approved models are: 18 Wp PV module with one 9 W compact fluorescent lamp (cfl), 37 Wp PV module with two 9 W cfl or one 9 W cfl and a fan, 74 Wp PV module with four 9 W cfl or two 9 W cfl and a fan/TV, [15]. Subsequently, the capital subsidy payable on the 74 Wp model has been curtailed to the level of the 37 Wp model, lowering the incentive thereon. Further, the Ministry has imposed price ceilings on systems, to be packaged with comprehensive maintenance contracts for 2, 5 or 10 years.

Over the tenure of the country's tenth five-year plan 2002–2003 through 2006–2007, (Indian fiscal year: April–March) against a target of 202,000 solar home systems (SHS), the program has helped deploy a total of 107,904 systems, corresponding to a 53% accomplishment of targets. The budget estimates and actual achievement of targets for each of the years, as reported by the Ministry are as laid out in Table 2.1

2.2.2 Solar Thermal Water Heating Systems

Solar thermal water heating systems (SWH) replace electric or LPG geysers in urban settings and firewood or other fuels, appliances and techniques used to heat

---

[5] http://mnre.gov.in.

Table 2.2 Year wise targets and installations of SWH systems

Year	2002–2003		2003–2004		2004–2005		2005–2006		2006–2007	
INR (crore)	11.00	9.73	12.00	9.90	14.00	6.91	50.00	24.89	45.75	13.23
SWH (area m²)	50,000	45,000	55,000	0	100,000	150,000	400,000	400,000	400,000	400,000

water in rural areas. The reduced consumption of such fuels as firewood, coal, furnace oil, etc., contributes to mitigation of carbon-dioxide emission and reduction in degradation of the environment. SWH have healthy monetary paybacks and with an LPG or electrical backup for non-sunny days, can provide heated water round the year. Large scale deployment of the SWH also helps reduce peak load demand for the electric utility [22].

The Ministry has encouraged a range of financial intermediaries (FI), including state-owned and private sector banks to participate in the low-interest ("soft-loan") scheme to extend loans of up to 5 years' tenure, not exceeding 85% of the project cost, at subsidized rates of interest. The Ministry compensates the implementing FI for the difference between prevailing commercial interest rates and the subsidized rates on offer. The subsidized rate of interest is available to domestic end-users, commercial establishments, cooperative societies and real estate developers [16]. Key attributes of the Solar Thermal Energy Program are: the absence of upper limits to the sizing of systems and the inclusion of the evacuated tubular collectors (ETC) under the soft-loan scheme.

In sharp contrast to the figures reported above in the context of solar home systems, over the same period, against a target of 1,005,000 m² of collector area, the program has contributed to the deployment of a total of 995,000 m², corresponding to a 99% accomplishment of targets. The budget estimates and actual achievement of targets for each of the past five years, as reported by the Ministry are as laid out in Table 2.2.

2.3 Quantitative Equivalence of the Two Subsidy Schemes

Chandrasekar and Kandpal [5] have computed the effective capital cost of solar energy technologies consequent to the fiscal and financial incentives provided by the government and have established indifference levels for various input parameters. For instance, they demonstrate that the provision of a low interest loan ("soft loan") at 2% annual rate of interest would be equivalent to a capital subsidy of 14.32% from the end-user's perspective.

In adopting an alternative view, in this section, subsidy provision is viewed from the government's perspective to help identify the indifference level between allocation towards the capital subsidy and the interest subsidy. Next, the *effective* cost to the end-user is computed using a larger set of input parameters, to establish the equivalence between the two schemes. This discussion leads to an analysis of the restrictions imposed by the capital subsidy program and the enlargement of choices achieved through the implementation of the interest subsidy scheme.

One could view the difference merely in the modalities of implementation, as opposed to the design of the two subsidy schemes itself. Yet, it should be brought out that a subsidy scheme *where choices are made by agencies other than the end-user concerned are regressive* and efforts need to be made to empower the consumer make an informed choice. It is often overlooked that end-users routinely make purchase decisions relating to televisions and mobile phones and associated services, and are, for some reason, considered incapable of deciding for themselves, the number and distribution of light points they should have in their own homes!

The following notation is used to highlight the equivalence and the differences between capital and interest subsidies.

C_o	Capital cost of system (currency units)	I_t	Interest amount payable at time t (currency units)
I_b	Interest rate set by the commercial bank (%)	P_t	Amount outstanding at time t (currency units)
I_s	Interest rate subsidy buydown offered by the government as promotional incentive (%)	R_t	Principal amount repaid in time t (currency units)
I_g	Discount rate for government funds such that $I_g < I_b$ (%)	C_{is}	Discounted value of the interest subsidy paid by the government (currency units)
I_m	Market interest rate paid by borrower to access the net-of-subsidy amounts such that $I_g < I_b < I_m$ (%)	CF_{is}	Periodic payments by the government towards interest subsidy on the loan (currency units)
I_w	Weighted average interest rate paid by borrower such that $I_b < I_w < I_m$ (%)	f_{cs}	Fraction of capital cost paid as subsidy (%)
F_{cs}	Portion of capital cost paid as subsidy (currency units)	f_{sl}	Fraction of capital cost available as soft loan (%)
PMT_t	Periodic payments including principal and interest on outstanding balance (currency units)	CF_w	Periodic payments by the end-user towards repayment of the loan (currency units)
		C_w	Discounted value of end-user loan service (currency units)

As shown in result 1 of annex 1 to this chapter, purely basing on the direct subsidy expenditure per system, the government is indifferent between offering an interest subsidy to lower the rate for the borrower by I_s and providing a capital subsidy F_s equal to $C_{is,}$ where,

$$C_{is} = \sum_{i=1}^{n} (CF_{is})/(1 + i_g)^t$$

Table 2.3 Input parameters
to compute the indifference
between capital and interest
subsidies

System cost	INR 20,000 (~ US$ 500)
Bank rate of interest	12%
Discount rate for government funds	6%
Market rate for end-user from informal and semi-formal sources	16%
Proportion of soft loan available	85%
Tenure of the loan	5 years

From the end-user's perspective, even as a capital subsidy of F_{cs} is made available, s/he is required to mobilize the net-of-subsidy amount $(C_o - F_{cs})$. This is mobilized at commercial rates prevailing in the markets, I_m, commensurate with the borrower's risk class. In the scenario where an interest subsidy is provided, the weighted average cost of the loan, I_w is given by:

$$I_w = (I_b - I_s)^* f_{sl} + I_m(1 - f_{sl})$$

The annuities payable by the end-user are computed and discounted at the prevailing market rate to arrive at the equivalent upfront payment, as perceived by her/him.

$$C_w = \sum_{i=1}^{n} (CF_w)/(1 + i_m)^t$$
$$= F_{cs}(\text{to establish the indifference level})$$

It is now possible to compute the indifference values for the government and simultaneously for the end-users, for a range of input values.

For the input parameters provided in Table 2.3 above, the interest buy down is 8.21% equivalent to a capital subsidy of INR 4,615 or about 23% of the capital cost. The net-of-subsidy amounts are mobilized by the end-user at the prevailing market rates. The similarity among the capital and interest subsidy schemes ends here.

2.4 Qualitative Divergence of the Two Subsidy Schemes

1. The principal difference between the schemes under discussion is the mode of delivery. The capital subsidy is delivered through nodal agencies and other qualified non-governmental organizations, while the interest subsidy is routed through the existing network of commercial banks and financial intermediaries. The not so insignificant indirect cost of routing the capital subsidy through parallel bureaucratic channels is not reflected in the computation above. In certain situations, as observed by industry watchers, subsidy management begins to justify the existence of the bureaucracy and reciprocally, enhances such entrenchment, simultaneously creating rent seeking opportunities.

2. The capital subsidy budgets are determined at the federal level and are drilled down to the state level implementing agencies by prescribing numbers of subsidy-eligible systems. The implementing agencies, further allocate, rather arbitrarily, quotas to select vendors. This leads to the creation of a supply-side dynamic where oligopolists market their wares in partially overlapping or non-overlapping intrastate markets, with fixed prices and identical product. In short, the oligopolists, almost inadvertently, end up acting as a monopolist. This is, however, a feature of the implementation process and not intrinsic to the design of the scheme itself.

3. Further, in order to standardize qualification for the capital subsidy, rigid eligibility criteria and technical specifications have been drawn up for the systems. This, by itself, curtails choice. For instance, in the above list of specified systems, a consumer is not permitted to acquire a 37Wp system capable of powering one lamp and one fan and to combine it with the 18Wp system to power an additional lamp. By restricting the modular expansion or contraction, the pre-designed quantity packages restrict the choices available to the end-users.

4. The one differentiator that could have emerged in this set-up is the quality of post-installation service, which is a function of the rigor of enforcement by the subsidy- implementing agency. It has been observed on several occasions that rural service providers scale down operations or completely go out of business upon the termination of capital subsidy programs, indicating that their businesses are only as sustainable as the subsidies themselves. The systems installed under the erstwhile programs, thus suffer from a lack of professional service provision for the remaining portion of their technical lives.

5. The net-of-subsidy price is low enough to eliminate all economic profits on non-subsidized systems, and hence deters entry and supply of systems with different technical specifications or the set up of service networks. This also ends up discouraging the provision of newer technology (evolving subsequent to the notification of technical specifications) or different component combinations that could help drive end-user prices down. It is virtually impossible for an individual in India to walk into a store, any kind of store, and walk out with a solar PV system of her/his, paid for in cash.

6. In contrast, the interest subsidy scheme on offer through commercial banks is not constrained by price, installation size or single borrower limits. The banks have demonstrated keen interest in offering the soft loans to promote the installation of the SWH. The evaluation criteria do not prevent the end-user from choosing a 5,000 liter-per-day (lpd) or a 10,000 lpd SWH over a 100 lpd installation. Further, the systems are installed and maintained by vendors whose systems and components are certified by the Bureau of Indian Standards (BIS).

7. Unlike the SHS vendors, the SWH vendors have aggressively invested in building brands and reputations through prompt and disciplined post-installation service. The quality of the system and the service, therefore, effectively sets vendors apart, who eventually earn premiums on such brand equity. This is not possible in a price constrained environment. In fact, at the most elementary

level, policy-makers should realize that price ceilings lead to shortages: in case of the system-service package, the intangible attributes such as (superior) installation-quality and prompt post sale service are found to be in acute short supply.

8. Further, the SWH vendors have also marketed evacuated tubular collectors (ETC), which represents a lower-priced technology variant, compared to the traditional flat-plate collectors. Being static in definition, the specifications for SHS do not allow for the inclusion of newer technologies viz., LED clusters to supplement the compact fluorescent lamps—as such designs would not be eligible to avail of the capital subsidy.

2.5 Choice and Inclusive Development

Arrow [1] asserts that a rational individual would choose an opportunity which helps achieve the highest level of utility, and that adding new opportunities cannot worsen the choice. Irrespective of whether the additional element is better than the originally chosen one, giving an individual more choice is per se valuable. In practical terms, the choice need not pertain to a range of goods: it would equally apply to quantities and qualities of a good supplied. Packaging goods in discrete quantities and pre-designed price–quality combinations adversely affects the poor. In most situations this is tantamount to fitting a solution to a problem, rather than the other way around. Quality standards prescribed by regulatory authorities could, thus, put goods out of reach of certain sections of society. Economic progress leads to the invention of higher quality products and this may include or exclude a larger proportion of the population, depending on concomitant changes in effective unit costs and price levels within the existing industry structure.

2.6 Subsidy Policy and the Range of Choices

In a market with a single quality of a good, firms compete exclusively through price strategies. In the SHS market, the quality and price are exogenously determined and hence the vendors behave as a cartel and barely compete. Where implementing-agency oversight and enforcement is weak, service support (and quality of the installation itself) is overlooked. The failure of a monopolist—or oligopolists acting in concert as in this case—to supply all the market and the prospect of a positive economic profit would attract entry. However, if the monopolist produces a superior product, as in the case of the specified solar home systems, while an entrant chooses to offer a product with a lower specification, at a lower price, the profit-maximizing response from the erstwhile monopolist would be to lower the price charged [2].

The level of profit for a monopolist supplying a commodity at price p, with a constant marginal cost c, per unit and a fixed cost of production k is given by

$$\Pi = (p - c)[1 - F(p/\lambda)] - k$$

where Atkinson [2] defines λ as a person's productivity-related, wage-influencing parameter. Further, λ_{max} and p_{max} are respectively, the quality and price of a superior (say prescribed technical specification) product and λ and p are the quality and price of a relative lower specification product, the market is segmented into consumers who can afford the superior product, those who can afford only the inferior product and those excluded. The pricing of the lower quality products by the same vendor needs to account for the revenue loss from customers who switch from the superior product ("cannibalization")—a common feature in the automotive industry, for instance. The profit from offering the two products is given by

$$\Pi = (p - c)[1 - F(w^*)] + (\rho - \gamma)[1 - F(w^{**})] - k - k_{max}$$

where $\rho = p_{max} - p$; $\gamma = c_{max} - c$ and w* is the threshold exclusion wage while w** empowers the purchase of the superior product. Since the two products in question are not perfect substitutes—though they may essentially perform the same function of lighting the household—the pricing strategy for the inferior product is considered independent of the superior product.

Suppose the incumbent supplies the superior product as prescribed under the capital subsidy scheme, a potential entrant could sell an inferior product at a lower price. An interest subsidy scheme can facilitate the purchase of such a system while the capital subsidy scheme accompanied by rigid technical specifications excludes it. The arguments posed by Atkinson [2] for two discrete commodity quality stipulations can thus be extended to cover a large number of variants assembled through mixing and matching quality certified components, maximizing choice and including a larger proportion of the population. Additionally, we relax the assumption relating to product indivisibility, to reflect the possibility of numerous combinations of system components.

Where the incumbent has the option to adjust price, a reduction in price by surrendering a portion of the monopoly rents is likely, in an attempt to ward off competition. If, on the other hand, the rival SHS is preferred to non-purchase, the entry leads to widening the range of systems and to reducing the income necessary to purchase each of the models available. Prices and availability of goods are determined endogenously by the decisions of suppliers faced with specified market conditions. Vendors supplying product under inflexible and exogenously imposed price-ceilings are tempted to compromise on the service component of the product-service package. As for the end-users, the monopoly price may exclude some customers at the lower rungs of society from the market and hence deprive them of the product.

If both, the superior and the inferior product are brought to the market by the monopolist—in this case the existing vendors—and given the assumption that the profit from the supply of both products is greater than that earned by supplying either product alone, the price of the inferior product as defined by Atkinson [2] is:

$$p = \frac{1}{2}(\lambda w_{max} + c)$$

The price for the said inferior product when offered by the entrant is given by:

$$p = \frac{1}{2}(\lambda p_{max}/\lambda_{max} + c)$$

which is lower by

$$\Delta p = \frac{1}{2}(\lambda w_{max} + c) - \frac{1}{2}(\lambda p_{max}/\lambda_{max} + c)$$
$$= \frac{1}{2}\lambda(w_{max} - p_{max}/\lambda_{max})$$

If the maximum wage endowment in the target market, w_{max}, and price per unit of quality for the superior commodity, p_{max}/λ_{max}, remain constant, the price reduction through liberalization of the market is a function of the quality chosen for the inferior product.[6] By relaxing the indivisibility assumption, we are able to fashion multiple variants at corresponding price points.

2.7 Sector, Segment and Market Neutrality

The European roof-top market has grown on the back of low-cost finance and high feed-in tariffs (FiT) offered and have helped advance an environmentally benign image for the power sector, especially in countries like Germany and Spain. Convergence with the utility grid in Japan has been driven by similarly structured subsidy programs (and comparatively high tariffs of grid-supplied power). In urban and sub-urban India, the PV systems would be filling the gap in the power supplied by the utility grid and would compete against other alternatives available in the urban context, viz., the ubiquitous diesel generators or inverter-based systems which draw power from the grid and typically use automotive batteries for storage. Most computers are connected through uninterruptible power supply systems (UPS), to bridge the gap during grid outages.

A diesel generator of equivalent capacity has a life cycle cost of almost twice that of a PV system, in some situations [21], viz, on island geographies, where transportation costs contribute substantially to the landed cost of fuels. Kolhe et al. [12] have worked on the economic viability of the stand-alone PV system as compared to the diesel-powered system and, given Indian operational parameters, have arrived at energy demand range of 15 kWh/day (worst case) to 68 kWh/day (best case) where PV-powered systems are cost competitive. Along similar lines, Hansen and Bower [10], opine that electricity planners have focused attention on

---

[6] In this context quality does not imply mediocre product or indifferent design, but lower wattage of lamps and the like.

lowering electricity costs at generation sites, while transmission and distribution losses and thefts have taken their toll. Further, subsidized tariffs on grid-supplied power reduce the incentive to look for alternatives. They believe that small scale generation technologies with power ranges as low as 5 kW (indicative range) are now in a position to compete economically with grid connected central generating capacity. This evolution in economics of alternatives opens up a world of possibilities for the production of electricity close to the point of consumption, thus reducing both technical and non-technical losses (power theft).

A rapid increase in volumes from a relatively "easy" market ("low hanging fruit") would help reduce prices for the market as a whole ("scale effects"). Additionally, higher installation density and a critical mass of systems help offset the higher costs of providing energy services in dispersed expanses ("network effects"). Appropriate banking channels could be identified to route the interest subsidy, as for instance, through rural branches of commercial banks, cooperative banks, etc. The policy intended for the promotion of a certain technology should be neutral to territorial demarcations, applications and classes of end-users.

2.8 Inferences

Financing mechanisms should be non-distortionary, inexpensive to administer and competitively neutral, enhancing allocative efficiency and not benefiting a few firms at the expense of others. Liberalized entry and competition encourage innovation and lead firms to generating a wealth of service, price and quality options to reach the target markets. The interest subsidy routed through banking channels is a superior option as it does not depend on quota or geographic allocation and merely requires that quality marked components be employed. Since the soft loan is linked to the capital cost of the installation, prepackaged system sizes and exogenously determined prices are not prerequisites. A competitive market enables innovation and the creation of brand equity, multiple price points and a range of robustly engineered quality options, while ensuring prompt and disciplined maintenance service for the systems. The end-users learn to discern product and service quality, over time. Additionally, the interest-subsidy schemes could be structured so as to ensure that product costs are progressively lowered through technology evolution (identical product at lower costs or superior product at identical costs), which would not be feasible under static capital-subsidy regimes.

The bureaucratic channels intent on controlling the distribution of capital subsidy could be put to better use in creating awareness among the target population and in encouraging vendors to use quality certified components. More generally, policy makers and their executive wings should abstain from replicating activities that could easily and more efficiently be carried out by non-state actors, specializing in such activities as banking, retailing, etc.

With the enlargement of the interest subsidy scheme, the systems, rather than the end-users are required to meet eligibility criteria. The consumer is at complete liberty to choose the vendor, the lender, the combination of sub-systems in the quality range between λ_{max} and λ and a price point between p and p_{max}, commensurate with his/her own endowment. At the end of it all, development is about empowering the masses and providing an enabling environment for people to choose for themselves.

2.9 Annex

$$PMT_t = C_o[i(1 + i_b)^n]/[(1 + i_b)^n - 1]$$
$$I_t = I_b * P_{(t-1)}$$
$$R_t = PMT_t - I_t$$
$$P_t = P_{(t-1)} - R_t$$
$$CF_{is} = I_t * P_t$$

Interest subsidy paid during the tenure of the loan is discounted at the rate of return on alternative use of government funds I_g

$$C_{is} = \sum_{i=1}^{n} (CF_{is})/(1 + i_g)^t$$

The equivalent capital subsidy as a proportion of system cost is given by C_{is}/C_o. The government is therefore indifferent between offering an interest subsidy to lower the rate paid by the borrower by I_s or a capital subsidy, F_s equal to C_{is}.

— Result 1.

Capital subsidy provided = F_s
Net-of-subsidy amount to be mobilized = $C_o (1 - f_s)$

$$C_w = \sum_{i=1}^{n} (CF_w)/(1 + i_m)^t = F_{cs}$$

— Result 2.

References

1. Arrow KJ (1995) A note on freedom and flexibility. In: Basu K, Pattanaik P, Suzumura K (eds) Choice, welfare and development. Oxford University Press, Oxford
2. Atkinson AB (1995) Capabilities, exclusion, and the supply of goods. In: Basu K, Pattanaik P, Suzumura K (eds) Choice, welfare and development. Oxford University Press, Oxford

3. Bhattacharyya SC (2007) Power sector reform in South Asia: Why slow and limited so far? Energy Policy 35(1):317–332
4. Carlos JC (1997) Photovoltaic technology and rural electrification in developing countries: the socio-economic dimension. IPTS Report No. 19, www.jrc.es/iptsreport/vol19/english/ENE1E196.htm
5. Chandrasekar B, Kandpal TC (2005) Effect of financial and fiscal incentives on the effective capital cost of solar energy technologies to the user. Sol Energy 78:147–156
6. Chattopadhyay P (2007) Testing viability of cross subsidy using time-variant price elasticities of industrial demand for electricity: Indian experience. Energy Policy 35(1):487–496
7. Clarke GRG, Wallsten SJ (2003) Universal service: empirical evidence on the provision of infrastructure services to rural and poor urban consumers. In: Brook PJ, Irwin TC (eds) Infrastructure for poor people. The World Bank, Washington
8. The Economist (2005) India's electricity reforms: underpowering. The Economist, 22nd September
9. GEF (2001) Results from the GEF Climate Change Program—Evaluation Report # 1-02. Monitoring and Evaluation Unit, Global Environment Facility
10. Hansen CJ, Bower J (2003) The political economy of electricity reform: a case study in Gujarat, India. Report by Oxford Institute for Energy Studies, p 47. http://econpapers.repec.org/RePEc:wpa:wuwpot:0401006
11. Jacobson S, Lauber V (2006) The politics and policy of energy system transformation—explaining the German diffusion of renewable energy technology. Energy Policy 34(3):256–276
12. Kolhe M, Kolhe S, Joshi JC (2002) Economic viability of stand-alone solar photovoltaic system in comparison with diesel-powered system for India. Energy Economics 24:155–165
13. Kumar A, Jain SK, Bansal NK (2003) Disseminating energy-efficient technologies: a case study of compact fluorescent lamps (CFLs) in India. Energy Policy 31:259–272
14. Makhijani A (2005) India should choose Iran, not US. www.rediff.com, 28 December 2005
15. MNRE (2005) Annual Report of the Ministry of New and Renewable Energy, 2004–2005
16. MNRE (2006) Annual Report of the Ministry of New and Renewable Energy, 2005–2006
17. Osterkorn M (2007) Solar thermal to score in SA? Renewable Energy Focus, pp 26–30
18. Pearce JM, Harris PJ (2007) Reducing greenhouse gas emissions by inducing energy conservation and distributed generation from elimination of electric utility customer charges. Energy Policy 35(12):6514–6525
19. Rader NA, III Short WP (1998) Competitive retail markets: tenuous ground for renewable energy. Electricity J 11(3):72–80
20. Radulovic V (2005) Are new institutional economics enough? Promoting photovoltaics in India's agricultural sector. Energy Policy 33(14):1883–1899
21. Sunderasan S (2006) Transforming solar thermal: policy support for the evolving solar water heating industry. Refocus 7:46–49
22. Taylor R, Abulfotuh F (1997) Photovoltaic electricity in Egypt. Renewables for sustainable village power. Project Brief, Department of Energy, NREL
23. World Bank (2006) New renewable energy: a review of the World Bank's assistance. Independent Evaluation Group
24. Yang H, Wang H, Yu H, Xi J, Cui R, Chen G (2003) Status of photovoltaic industry in China. Energy Policy 31:703–707

Chapter 3
Economic Rents and the Power of Scarcity

"It is commonly said, that a sugar planter expects that the rum and the molasses should defray the whole expense of his cultivation and that his sugar should be all clear profit. If this be true, for I pretend not to affirm it, it is as if a corn farmer expected to defray the expense of his cultivation with the chaff and the straw, and that the grain should be all clear profit."

Adam Smith

"Rent of Land", The Wealth of Nations.

The previous chapter dealt with the power of choice and concluded that development policy should ensure that the end-users are adequately informed and empowered to make choices for themselves. In the medium-term, fiscal and financial incentives should be geared towards bringing about sufficient awareness and in supporting, rather than supplanting, existing distribution channels for technology options or for financing. In a similar vein, the next chapter analyzes the power of scarcity and highlights the backward drift of surpluses to the owners of the scarce resources: sugarcane farmers in the case of the sugar industry. Even as the sugar industry is studied in detail, the central theme pertaining to scarcity rents is equally applicable to the solar PV industry (silicon wafers), wind energy sector (turbines and sites), small hydro sector (hydro-mechanical equipment and sites) etc.

Resource scarcity rents embody multiple elements dictated by economic theory and intuition [35] and such values are often benchmarked against plausible alternative applications—opportunity costs—of the resource in question [17]. In industries seeking to exploit scarce resources viz., telecom (electromagnetic spectrum), aviation (airport capacities) and petroleum leases, auctions are preferred ahead of administrative decisions to enhance efficiency in capturing resource rents [40]. Yet, the distribution of such economic rents among the host nations, the developer corporations and consumers tends to vary across industry

S. Sunderasan, *Rational Exuberance for Renewable Energy*,
Green Energy and Technology, DOI: 10.1007/978-0-85729-212-4_3,
© Springer-Verlag London Limited 2011

sectors. For example, while it is observed that in petroleum and tin, exporting nations capture a significant proportion of the embedded rent [20], in the case of non-fuel primary commodities, agencies involved in downstream activities tend more frequently, to retain such value [12], possibly owing to the specialized skills and processes involved. The sharing of profits and rents, therefore, is dictated by the relative scarcity of factors going into production and the availability of such information in the public domain that could help spread economic profits more evenly.

Land, and frequently products emanating from land, like other scarce inputs earn 'what they can' as earnings on land are price-determined, rather than price-determining, entirely basing on the value of the finished product. The price of the finished product emanating from land serves as a signal to help equilibrate investments in material supply—crops grown or mineral extracted—and processing on the one hand, and market demand on the other. The absence of pricing transparency, combined with the complexity of supply mechanisms and the distribution of power among agencies involved, could lead to unsustainable extraction of resources, industrial overcapacities and other such imbalances, thereby, potentially challenging the viability of private sector investments, and thereby discouraging further investments into various stages of the value chain [33].

3.1 The Indian Sugar Industry: An In-depth Case Analysis

Globally, demand for sugar tends to be inelastic, considering that there is no substitute product in the short-run (est. elasticity = −0.31) and little potential for substitution in the long run[1] (est. elasticity = −0.45) and consumption tends to rise in tandem with economic growth [9, 15].

The Indian sugar industry is a unique blend of aggression contributed by the private sector, passivism of the moribund state-owned and cooperative sectors, the socio-political realities of interacting with large numbers of farmers, and the rapidly evolving global demand–supply situation. That cane is required to be crushed within 24 h of being harvested leaves the farmer as well as the millers vulnerable, and keeping the farmers engaged in order to ensure timely and reasonably-priced supplies is critical to sustaining mill operations. Each sugar mill sources cane from farmers within a captive area covering a 15 km radius and advises them on cultivation. Select industry participants have called for abandoning such reservations, while simultaneously, others have called for enlarging the captive area to suit the crushing capacity of larger mills. Further, considering that a typical 5,000 t of

---

[1] High-fructose corn syrup (HFCSF)—called isoglucose in the UK and glucose-fructose in Canada—is typically used as a sugar substitute in processed foods and beverages, including soft drinks, yogurt, industrial bread, cookies, salad dressing, and tomato soup. Its potential as a substitute for sugar is expected to be acutely limited. (http://en.wikipedia.org/wiki/High-fructose_corn_syrup).

sugarcane crushed per day ("5,000 TCD") capacity sized mill is required to interact with about 50,000 farmers, backward integration into cane cultivation or contract-farming is not a viable option in the medium-term[2] [21].

Demand elasticity is partly determined by the industry structure, comprising concentration, the competitive behavior among firms, protection from domestic and foreign entry and potential for product innovation [32]. Sugar, being a generic, homogenous commodity with a large weight-to-value metric, does not readily lend itself to product differentiation and innovation, while transportation costs add significantly to product pricing when exported over long distances. Additionally, in the Indian setting, individual mills tend to be too small to influence global pricing for the commodity. Volatility in pricing, therefore, stems almost exclusively from supply side disruptions and is a consequence of concentration of production and export in a few countries viz., Cuba, Brazil, Thailand and Australia. Free market supply is also impacted by the diversion of cane to ethanol production in countries like Brazil [37].

In India, state administered support prices for cane and market-determined prices for sugar[3] keep the industry dynamic finely balanced. With no control on the principal input required, and exposed to the commodity cycle, sugar mills have chosen to enlarge crushing capacity to derive scale-economies while also diversifying into ethanol production and by investing in cogeneration of electric power (also referred to combined heat and power "CHP")—the simultaneous generation of steam and electric power which is fed to the utility grid (thus converting sugar mills into "sugar-energy-complexes").

3.1.1 Cogeneration in Indian Sugar Industry: Theoretical Potential and Practical Hurdles

The cogeneration potential of the Indian sugar industry is estimated at over 5000 MW, growing to over 10,000 MW with improved technology and additional resources. From a macroeconomic perspective, power plant investment in the sugar industry offers a realizable bridging of the demand–supply gap in power supply in the country. Sugarcane cogeneration is found to be technically and economically feasible if the generated electricity is priced at the long-term marginal cost [25].

Sugarcane cogeneration through combustion of bagasse—moisturized fiber residue from crushing cane, otherwise a mill waste—is considered

---

[2] This is in sharp contrast with Brazil, the largest cane/sugar producer in the world, where 75% of the sugarcane is grown by the mills themselves and only 25% by independent producers [23].

[3] A portion of sugar produced (presently 10%, proposed to be raised to 20%) is procured by the government at administered prices for sale through the public distribution system. With a view to spreading supply evenly, the residual output destined for free market sale is released each month at 1/12th of the annual production.

environmentally benign as carbon dioxide sequestered by sugarcane plants is more than that released during combustion. This helps the sugar mills earn additional revenues from the sale of carbon emission reduction certificates (certified emission reductions "CER"); frequently, the additional revenues from the sale of power and CER are believed to help sustain or even turn-around an otherwise unviable sugar operation [7].

The tariff payable on the power so generated and exported to the grid varies from state to state, ranging from INR 2.63 per kW h to almost INR 4.0 per kW h, with vast differences in the utility imposed charge for wheeling and banking of power. Other incentives offered include capital subsidies and cost-sharing for transmission lines in select states and duty incentives, accelerated depreciation on the equipment, tax holidays and other fiscal concessions from the federal government [29].

The theoretical annual CER potential from bagasse cogeneration in the 566 Indian Sugar mills is estimated at 28 million tons while observers have concluded that more favorable policies than presently implemented would be required to fully exploit the stated potential in fewer than 20 years [34]. For instance, the Electricity Act (2003) provides for "open access", with a rider though, that the same could be withheld during extraordinary situations and in public interest, an option that most state governments choose to exercise citing the persistent power deficit. Facing a substantial shortfall in power generation especially from hydroelectric plants, the southern Indian state of Karnataka raised the tariff for cogeneration plants to INR 7.24 [7], but imposed restrictions on the sale of power to third parties. While the higher tariffs when offered are attractive, the same need to be tempered with the probability of timely realization of proceeds from the sale of power, given the uncertain financial health of the state-run electricity transmission companies. Considering the high and growing power deficit, power sector regulators are contemplating the imposition of a cap on power tariffs charged by 'merchant power producers'—electricity plants that do not execute long-term power purchase agreements [27]. Inter-state sale of power is yet to be fully liberalized and power trading is still in its infancy.

From the sugar millers' perspective, the greatest challenge however, is that world sugar prices are relatively volatile while cane prices are administered—managed by elected representatives and bureaucrats—and have progressively been on the rise; lowering sugarcane price is not a politically viable option for policy makers; diversifying into power generation is not liberating either as both power prices and sugarcane prices are administered and each could be altered almost unilaterally.

3.2 Determination of Prices and Testing for Granger Causality

Given the competitive structure of the Indian sugar industry, and the generic nature of sugar, one would anticipate that each individual mill is a 'price-taker' and that this price itself is discovered basing on the marginal costs of production, including among other things, the marginal cost of sugarcane paid to the farmers (cost-plus approach). Conversely, the sugarcane prices themselves could be determined working backwards from finished product prices, since sugar prices are market determined, and that cane prices should mirror movements in sugar prices (backward induction).

Testing causality [14] involves using F tests to ascertain whether the time-lagged information on variable Y provides any statistically significant information about a variable X in the presence of lagged X. If not, it leads to the conclusion that "Y does not Granger-cause X". The test also provides an indication of the time taken for the transmission mechanism to unravel.

A particular lag length 'p' is assumed to specify the autoregressive bivariate vector, and to estimate the following unrestricted equation by ordinary least squares (OLS):

$$x_t = c_1 + \sum_{i=1}^{p} \alpha_i x_{t-i} + \sum_{i=1}^{p} \beta_i y_{t-i} + u_t$$

$$H_0 : \beta_1 = \beta_2 = \cdots = \beta_p = 0$$

An F test of the null hypothesis is then conducted by estimating the following restricted equation also by OLS:

$$x_t = c_t + \sum_{t=1}^{P} \gamma_i x_{t-i} + e_t$$

Their respective sum of squared residuals are compared.

$$RSS_1 = \sum_{t=1}^{T} \hat{u}_t^2 \quad RSS_0 = \sum_{t=1}^{T} \hat{e}_t^2$$

If the test statistic

$$S_1 = \frac{(RSS_0 - RSS_1)/p}{RSS_1/(T - 2p - 1)} \sim F_{p,\, T-2p-1}$$

is greater than the specified critical value, then the null hypothesis that Y does not Granger-cause X is rejected.

The Granger-causaltiy test has been applied to assess causality in diverse situations. Hoffmann et al. [18] employ the Granger causality test between foreign direct investment (FDI) and pollution levels in the host country and observe that higher pollution levels Granger-cause inward FDI flows in low-income countries, while the causality is reversed in middle-income countries. Liu et al. [24] employ the Granger causality test to identify the causal relationship between economic growth and public expenditure and identify a unidirectional Granger-causality between federal outlays and change in GDP and not vice versa. Ghosh [10] confirms the existence of a unidirectional Granger causality leading from economic growth in India, measured by GDP per capita, to electricity consumption per capita in the country. Kilima [22] has observed a unidirectional Granger-causal

Table 3.1 The causality test of sugar cane and sugar

	Sugar → Sugar cane	Sugar cane → Sugar
Lag (weeks)	F test (p-value)	F test (p-value)
09	0.37951 (0.94487)	1.19876 (0.29348)
12	0.30445 (0.98857)	1.14737 (0.31934)
15	0.26093 (0.99789)	1.19439 (0.27212)
16	0.25320 (0.99873)	1.19858 (0.26498)
17	0.25208 (0.99911)	1.11801 (0.33289)
35	0.35538 (0.99979)	0.97666 (0.50973)
38	0.36356 (0.99985)	0.97949 (0.50790)
39	0.36362 (0.99987)	0.96097 (0.54042)
40	0.37057 (0.99986)	0.93509 (0.58660)
65	0.40416 (0.99999)	0.85158 (0.78069)
104	0.47602 (0.99998)	0.65315 (0.99166)

95% statistical significance 99% statistical significance

relationship between international prices for sugar and a few other agricultural commodities and corresponding local prices in Tanzania and that some of the shocks from the world market passed through to Tanzania, but not vice versa.

Wholesale price data (index 1993–1994 = 100) for sugar and electricity (final products) and sugarcane (input) are collected from the webpage of the Economic Advisor to the Ministry of Industry and Commerce[4] to test for causality for the period January 2000 through August 2009.

The tests are run for time lags ranging from 1 to 60 weeks and intermittently thereafter, and noteworthy results are presented in the respective tables. Bagasse, molasses and other intermediate products are not considered for analysis here, as they find diverse applications and their pricing is slated to influence or be influenced by the prices of respective finished products.

Sugarcane prices are reset each year, towards the commencement of the Indian "sugar year", in the month of October or thereabouts and more frequently, if found necessary. Intuitively, cane prices should be set at levels that would make it worth the while for farmers to cultivate cane and keep them from switching crops, and yet leave competitive margins for the sugar mills. The farmer's decision to grow cane is therefore co-determined by the relative viability of potential substitute corps, viz., paddy and pulses, discounted by switching costs, if any [1].[5] The results from the analysis of sugar and cane pricing listed in Table 3.1 point in no specific direction: cane prices influencing the determination of sugar prices ("cost plus") or sugar prices effecting changes in cane pricing ("backward induction"); neither Granger-causality is statistically significant (at the 99% or the 95% levels).

---

[4] http://eaindustry.nic.in/; Economic Advisor to the Ministry of Commerce and Industry, Government of India.

[5] Likewise, the wage rates for temporary labor and the very availability of such labor, are determined by alternative employment opportunities, including by the National Rural Employment Guarantee Scheme (nrega.nic.in).

Table 3.2 The causality test of sugar cane and electricity

	Sugar cane → Electricity	Electricity → Sugar cane
Lag (weeks)	F test (p-value)	F test (p-value)
21	0.41006 (0.99131)	1.72271 (0.02480)[*]
22	0.41487 (0.99200)	1.62846 (0.03680)[*]
34	0.46278 (0.99619)	2.36846 (0.000044)[**]
35	0.43643 (0.99811)	2.28739 (0.000075)[**]
36	0.41973 (0.99890)	2.19338 (0.00015)[**]
37	0.43442 (0.99861)	2.10988 (0.00027)[**]
38	0.41871 (0.99919)	2.03623 (0.00045)[**]
39	0.43049 (0.99903)	1.95104 (0.00085)[**]
40	0.42163 (0.99933)	1.89612 (0.00124)[**]
41	0.47318 (0.99780)	1.82263 (0.00215)[**]

[*] 95% statistical significance [**] 99% statistical significance

This may well be considered a reflection of reality. In what appears rather counter-intuitive, there have been instances when the support price for sugarcane is pegged at levels higher than the price of sugar it yields, [2].

The analysis of wholesale prices for sugarcane and electricity (Table 3.2) indicate a statistically significant unidirectional G-causality from electricity to sugarcane indicating that periodical adjustment of cane prices is strongly influenced by the wholesale price of electricity. A statistical result that is consistent with the sugar mills' diversification strategies and one that reinforces the quip that sugar is merely a byproduct in a sugar mill.

The economic analysis of the situation is far more compelling: wholesale electricity prices are determined by statutory bodies, namely the electricity regulatory commissions, staffed by independent experts and professionals. The incentive packages are designed to remove barriers and to encourage sugar mills to set up power generating units that help bridge the electricity demand–supply gap in an environmentally sustainable way, while helping the mills stabilize their own financial position. However, by working backwards from the power tariffs, it is possible for the political establishment to transfer such enhanced returns to the farmers through the administered prices of sugarcane. The sugar mills therefore get to retain the incentives for a period of about 34 weeks, i.e. approximately 1 year's worth of sugar production, after which, incremental earnings are transferred to the sugarcane farmers as an economic rent.

3.3 The Prisoners' Dilemma Model

An iterated prisoner's dilemma model, with outside options yielding stochastic payoffs, is formulated to help arrive at the equilibrium for the sugar cogeneration industry. It is noted that in such a game, more frequently, the decision to switch states is driven by inertia rather than the urge to respond to the counter-party's

Table 3.3 The prisoners' dilemma payoff matrix

Sugar mill (M)	Cane farmer (F)		
		Cooperate (C)	Defect (D)
	Cooperate (C)	R_m, R_f	S_m, T_f
	Defect (D)	T_m, S_f	P_m, P_f

and $T_i > R_i > P_i > S_i$ and $(T_i + S_i) < 2 R_i$ for $i \in \{m, f\}$

move [36]. In a given situation, it is entirely possible that one of the players is compelled to be more "inertial" than the other. In the present case, the sugar mill, having made the sizable investment, is required to stay in the game for up to 20 years or more, to recover the first costs and to reward shareholders appropriately. Cogeneration and ethanol production help diversify sources of revenue but still require sugarcane as the principal input and diversification and additional investments only serve to enhance dependence on cane and consequently, the mill's "inertia" with respect to staying in the game.

On the other hand, the farmers make the "cooperate/defect" decisions each season—on crop choice at the time of planting and on choosing to sell to the sugar mill on harvest.[6] This difference in planning horizons results in asymmetry in payoffs at the end of each sub-game. Such asymmetry is also induced, in part, by government action, helping the farmers capture scarcity rents through periodic revision in cane pricing, *effected independent of ruling sugar prices*. The sustainability of mutual cooperation is clearly affected by the asymmetry of payoffs from each sub-game, and by the structure of the outside options available [8]. In general, higher the magnitude of gains from mutual cooperation relative to the payoffs from mutual defection, higher is the rate of cooperation; i.e., some players choose to cooperate in order to achieve the outcome that maximizes the *joint* payoffs from mutual cooperation.

Each season the sub-game level interaction between the sugar mill and the cane farmer is driven by individual payoffs from mutual cooperation or defection but also by the evolution in the risk appetite of the players, predominantly the sugar mill. For instance, when the 'sugar energy complex' commences operations, and the proportion of project debt in the capital structure is high, the mill would like to ensure predictable and stable cash-flows from operations. However, progressively, with reducing levels of project debt, the mill could afford to be less risk-averse and could choose to defect occasionally—by delaying payments for cane supplied, for instance. Using the standard notation[7] for prisoner's dilemma payoffs, the sub-game payoff matrix is modeled as shown in Table 3.3.

---

[6] On average, 1 in 4 farmers chooses to supply cane (25% of the cane produced) to the informal sector for *jaggery* and *khandsari* (cottage sugar) production [38].

[7] Mnemonically, these letters represent R(eward for cooperation), T(emptation to defect), P(unishment) and S(ucker) payoffs.

3.3.1 Stage 1

a. The 'sugar energy complex' is newly constructed and is debt-laden and hence highly risk-averse.
b. The farmers are informed of the support price on sugarcane announced for the planting season.
c. Prevailing spot and futures prices for potential substitutes viz., paddy, wheat, pulses etc. are also known.
d. There is no scope for cane-poaching by competitor mills from within the (15–25 km radius) captive area assigned to the sugar mill and adequate cane is available to run the entire length of the crushing season (between 6 and 10 months).
e. Individual farmers make decisions based on the relative viability of output.

At the launch of the asymmetric iterative game, the 'disadvantaged' player, the mill, makes the first move by making the investment in the mill. Defection is not an option for the mill for several seasons to come. The farmer finds the cane prices attractive relative to substitute crops and chooses to 'cooperate' by growing cane. On harvest day, the farmer is required to choose between supplying to the mill and to alternative buyers, viz., *jaggery* makers (or other informal sector players), who offer higher prices but with patchy payment records. The farmer chooses to 'cooperate' again, i.e., supply to the sugar mill, trusting that the mill would pay the promised amount on time. It should be noted that the probability that the farmer's choice to supply to the mill (or to the informal sector) is based on his/her expectation about how the game would be played if he/she chooses to play the game. The cane farmer's decision tree is depicted in Fig. 3.1.

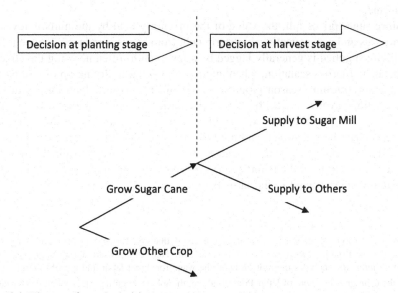

Fig. 3.1 The cane farmer's decision tree

The mill pays the farmers as promised. The payoffs from (C, C) in the stage-1 sub-game are respectively (R_m, R_f) the return on investment for the mill emanating from the sale of sugar, ethanol and power; and the remuneration for the farmer[8] growing sugarcane.

3.3.2 Stage 2

a. The 'sugar energy complex' is relatively new and debt-laden and hence continues to be risk-averse.
b. The support price on sugarcane is increased by the government.
c. Prevailing spot and futures prices for potential substitutes viz., paddy, wheat, pulses etc. are also known.
d. There is no scope for cane-poaching by competitor mills from within the (15–25 km radius) captive area assigned to the sugar mill and adequate cane is available to run the entire length of the crushing season.
e. Individual farmers make decisions based on the relative viability of output.

A large number of farmers enter sugarcane cultivation in response to the mill's 'cooperation' in stage-1, and the higher prices announced by the government. As the season progresses, there is a glut of cane. Eventually, the price of sugar falls in tandem with increased supply. The payoffs from (C,C) in the first part of the stage-2 sub-game are respectively (R_m, R_f) while the return on investment for the mill 'R_m' is lower, owing to lower *average* sugar prices and due to excess supply of ethanol/molasses, even as the quantity of power generated is higher; and a higher (compared to stage-1) remuneration 'R_f' for the farmer growing sugarcane.

Since sugar prices fall, the value of the inventory held by the mills is revised downwards and consequently, short-term debt from banks is reduced. Payment for electricity supplied is generally lagged by a few months often negating the effects of the annual tariff escalation, when applicable. Owing to drying up of free cash-flows as the crushing season progresses, the mill begins to default on payments. The resulting payoffs (T_m, S_f) from restricted cash-flows leave neither player better off, as $T_m (> R_m)$ in this context could be a misnomer as default by the mill need not necessarily be intentional and the farmers would have to be compensated for the delay.[9] However, $T_m > R_m$ if, and only if, adequate cash is available with the mill, and the return on the (short-term) investment for the mill is greater than the quantum of interest paid to the farmers.

---

[8] For the purpose of generalization, it is presumed that the farmer, as referred to in this discussion, is also the owner of the land being cultivated: compensation for supply of cane therefore *inter alia* includes the rent on land, the wage for direct labor and a profit margin.

[9] In the cane growing state of Uttar Pradesh alone, there have been instances when the farmers are cumulatively owed in excess of US$ 600 million [21].

Table 3.4 The Prisoner's dilemma payoff matrix with endogenous transfers

Sugar mill (M)	Cane farmer (F)		
		Cooperate (C)	Defect (D)
	Cooperate (C)	$R_m - H^*, R_f + H^*$	$S_m - H^*, T_f + H^*$
	Defect (D)	T_m, S_f	P_m, P_f

3.3.3 Stage 3

a. The 'sugar energy complex' renegotiates its project debts and intentional default is still not an option.
b. The support price on sugarcane is increased by the government since power output has grown in Stage-2.
c. Prevailing spot and futures prices for potential substitutes viz., paddy, wheat, pulses etc. are also known.
d. Competitor mills are set up bordering on the captive area of the existing mill.
e. Individual farmers make decisions based on the relative viability of output.

In a tit-for-tat response, a large number of farmers defect either by switching crops or by supplying cane to other players. Those who choose to terminate the game opt for the (C, D) strategy with corresponding payoffs. Cane availability is reduced in Stage-3 in response to the mill's 'defection' in stage-2. Simultaneously, competitor mills poach cane grown in the captive area of the existing mill. Eventually, the price of sugar rises in tandem with reduced supply.

The mills then commence an informal auction for the available cane. Each mill makes a binding pre-play offer to the farmer to induce cooperation. This routinely takes the form of a premium paid over and above the state-advised price for cane, guarantees against third-party debt etc. Since the statutory authorities employ the median firm as the benchmark to determine prices, relatively more efficient firms are better positioned to offer higher premiums. "A natural solution concept is subgame-perfect equilibrium; while one [the mill] wishes to offer enough to induce the other [farmer] to cooperate, it is best to offer the minimum amount that is required to achieve this goal" [6].

As shown in Table 3.4, the payoffs from (C, C) in the stage-3 sub-game are respectively $(R_m - H^*, R_f + H^*)$ where H* is additional compensation paid for cooperation. In effect, a portion of the additional returns earned by the mill (from higher sugar prices) is transferred to the farmer in the manner of scarcity rents. This transfer provides a "formalization of bargaining" and is not at the behest of a regulator or planner but is market driven, leading to an efficient outcome.

3.3.4 Stage 4

a. The 'sugar energy complex' is freed from debt service and occasional default to pick up 'T_m' is now an option.

b. The support price on sugarcane is increased by the government.
c. Prevailing spot and futures prices for potential substitutes viz., paddy, wheat, pulses etc. are also known.
d. Competitor mills set up bordering on the captive area of the existing mill are operational.
e. Individual farmers make decisions based on the relative viability of output.

A large number of farmers enter sugarcane cultivation in response to the mill's 'cooperation' in stage-3 and the higher prices announced. As the season progresses, there is a glut of cane. Eventually, the price of sugar falls in tandem with increased supply and mills investigate possibilities for export. The payoffs from (C, C) in the first part of the stage-4 sub-game are respectively (R_m, R_f) while the return on investment for the mill is lower owing to lower sugar prices and due to excess supply of ethanol/molasses, even as the quantity of power generated is higher; and a higher (compared to stage-3, but without the premium) remuneration for the farmer growing sugarcane.

The mill could choose to defect and hence earn the (T_m, S_f) payoffs. This decision, as with iterated prisoner's dilemma games in general, is driven by the discount rate faced by the mill. A high discount rate would lower the present value of future earnings. If the mill is under pressure to deliver high investor returns (as, for instance, with offering private equity investors an exit), a one-time realization of 'T_m' might serve its interests. The sucker's payoffs in the context emanate from the liquidation of the mill's assets, including by acquisition of the mill by the competition.

3.4 Exceptions to the Model—Sugar Cooperatives

The incentive structures faced by the farmer-members of cooperative societies are substantially different. The Indian state of Maharashtra, for instance, has actively supported the formation of cooperatives with a view to preempting potential exploitation of the growers by privately owned sugar mills. Like other sugar mills in the country, each cooperative enjoys monopsony power over its captive cultivation area covering a fixed radius around the factory. A majority of the farmer-members of these sugar cooperatives are small farmers while the cooperative societies themselves are generally governed by the larger farmers, who possess the wherewithal to negotiate government procedures involved in setting up the cooperatives first and subsequently, the political acumen required to retain power. In several situations, the cooperatives serve as stepping stones towards realizing higher political aspirations.

There is a substantial gap between the physical potential of bagasse based cogeneration and the practically realizable potential, especially in states like Maharashtra. Barriers to adoption of cogeneration in cooperative mills are largely financial and institutional, rather than technical. The "un-corporate" and highly politicized culture of the sugar cooperatives, poor financial health of the sugar mills (since surpluses are transferred among the farmers via high cane prices or otherwise), uncertainty relating to the financial health and credit-worthiness of the state electricity boards and their successors and an emerging policy environment deter banks and financial institutions from lending to such projects [16].

The politicization and rent seeking behavior is deeply entrenched among the cooperative societies, to the extent that the government has remarked that "the cooperative sector is proving to be against the spirit of competitiveness and professionalism" [13]. While the state advised price for sugarcane serves as a floor, passing-through the mill generated profits to the grower-members by way of higher cane prices could be viewed as being tax efficient in certain circumstances. Rent seeking behavior among the governing elite could skew such sharing and prevent equitable distribution. Since cooperatives are prohibited from distributing dividends by law, the surpluses are siphoned off by awarding contracts for construction of roads and other public amenities to companies in which the governing elite have an interest. "Moreover, to the extent that these public goods benefit people who are not sugar farmers, being associated with them comes with a substantial political advantage, which the politically ambitious larger farmers must value," [3].

The reallocation of costs and profits is also achieved through over-invoicing for complimentary inputs employed in sugar production or through setting up of downstream operations viz., distilleries. The median farmer-member being almost totally disenfranchised from the cooperative society is likely to have little control on the quantum and timing of the payments made for cane. Among other things, the cooperatives are known to default on their debts from banks and other financial institutions with relative impunity. Power cogeneration and other potential downstream operations are unlikely to be governed more transparently and the game theory model developed above for the private sector sugar mill would have to be adapted suitably for the cooperative setting, if at all.

3.5 A Model to Eliminate Reciprocal Defection

Each sugar mill receives its supplies of sugarcane from within a catchment area spanning a radius of 15–100 km on occasion, limited only by the logistical constraints associated with transporting the cane to the mill within 24 h of being harvested. Over time, the farmers gain familiarity with mill personnel and prefer to be locked into a 'reciprocal monopoly' with the mill. The mill is built to a defined design-crushing-capacity and the tenure of sugar production over the course of the year is limited by cane availability.

Cane production yields constant returns to scale, since soil conditions and irrigation tend to remain constant within the catchment area. There are 'N' farmers located within the catchment area, and 'n' farmers growing sugarcane (for $n \leq N$), basing on the relative attractiveness of cane and the expectation of prompt payment from the mill. If the average land holding in the area is 'm' acres, the total command area is '$N \times m$' and area under cane cultivation is '$n \times m$'.

The support price for sugarcane 'p_t' and the market price of sugar is p_t^*, are exogenously determined and known at the time of planting. The average farmer grows 'q' units of cane per acre which yield 'q^*' units of sugar (where q^* $\sim$9–12% of q) and earns $m \times q \times p_t$ as gross revenues and $\pi(p_t)$ the resulting value of profits. We define $\pi^*(p_t)/(m \times q \times p_t)$ as the minimum acceptable return for an individual farmer. A farmer decides on growing sugarcane if $\pi_k(p_t) > \pi_k^*(p_t)$, for $k = 1$ to n.

The capital cost of setting up a sugar mill is $G(K)$, and 'i' the weighted average cost of capital, leading to a service cost of $i_t^*G(K)$ in year 't'. If 'Q' is the amount of sugar produced by the sugar mill, gross revenues from the sale of sugar are $p_t^* \times Q$.

Additional revenues are generated from the sale of power, $e \times \tau_t$ where 'e' is the quantity of electricity generated and τ_t is the tariff applicable in year 't'. The sale of molasses, ethanol etc., yield revenues 'S_t' constrained by the prices of their respective finished products viz., alcoholic beverages for molasses or by the substitute products viz., petrol in the case of ethanol.

A large proportion of these revenues are paid out for the sugarcane delivered i.e. $m \times n \times q \times p_t$. Part of the revenues go towards crushing and other variable costs (including applicable taxes) $(c(K)Q)$ and for fixed capacity costs $F(K)$ and as cost of capital $i^*G(K)$.

Earnings (from sugar alone), available for appropriation among the suppliers of capital are then equal to:

$$i_t^*G(K) \equiv p_t^* \times Q - [m \times n \times q \times p_t + c(K)Q] - F(K) \qquad (3.1)$$

If i_t^{**} is the minimum acceptable return on capital employed so as to help service the total debt and to provide a competitive return on equity capital, the mill continues to operate for all $i^* > i^{**}$. The pricing decision is required to balance the demands of the farmers against the economic viability of the sugar mill by maximizing the welfare function:

$$\mathbf{W} = i^*G(K) + \pi(p) \qquad (3.2)$$

subject to $\pi(p) > \pi^*(p)$ and $i^* > i^{**}$, respective threshold values.

Ceteris paribus, in the Stage-2 subgame, an increase in 'p_t' reduces the mill's capacity to service capital employed by '$m \times n \times q \times \delta p_t$', and can be sustained only insofar as $i^* \geq i^{**}$.

$$i_t^*G(K)' \equiv -m \times n \times q \times \delta p_t$$

An increase in the number of farmers growing cane, consequent to the increase in support price results in enhanced sugar output and hence lower market prices for sugar. Revenues from the sale of sugar fall by '$Q \times \delta p_t^*$' and expenses increase by '$m \times \delta n \times q \times (p + \delta p_t)$'.

$$i_t^* G(K) \equiv (p_t^* - \delta p_t^*) \times Q - [m \times (n + \delta n) \times q \times (p_t + \delta p_t) + c(K)Q] - F(K)$$

The mill begins to default when i^* begins to dip below i^{**}. In the Stage-3 subgame, the farmers switch to other crops, viz., paddy, leaving fewer farmers to cultivate sugarcane. The lower aggregate availability of cane also attracts mills in the vicinity, disrupting the existing reciprocal monopoly. This triggers a bidding war for the available cane, resulting in unilateral transfers from the mills to the farmers. In effect, the higher sugar prices are transferred to those farmers who continue to cultivate sugarcane.

The mill manages to retain the loyalty of farmers through transfer payments H*, resulting from the auction, such that:

$$
\begin{aligned}
i_t^* G(K) &\equiv [(p_t^* + \delta p_t^{**}) \times Q - m \times (n - \delta n) \times q \times (p_t) - c(K)Q] - F(K) \\
&\quad - H^* \geq I^{**} G(K); \quad \text{and } \pi_k(p_t) + h^* \geq \pi_k^*(p_t) \quad \text{for all } k \in \{1, n\} \quad (3.3)
\end{aligned}
$$

where 'δp_t^{**}' is the average increase in the market price of sugar owing to an anticipated shortage, h^* is the individual farmer's share of H^* ($\Sigma h^* = H^*$) and 'δn' is the number of farmers who have 'defected' by growing other crops.

3.6 The Effectiveness of Product Diversification

Volatility in cane supply could be caused by natural or man-made factors, resulting in a few seasons of glut in production, followed by years of low production. Ironically, while low cane production could result in higher sugar prices, these gains are very substantially, passed on to the farmers to ensure continuity in supply. Experts have called for a fair and transparent pricing mechanism for cane, based, among other things, on sucrose content, as a plausible solution to tackling such volatility [31]. Additionally, it is observed that such volatility is enhanced by competition among mills located in close proximity.

To combat the high sugarcane prices on the one hand and the commodity cycle of sugar, on the other, the industry has resorted to diversification of the finished product, away from sugar and into power and ethanol. Consequently, the sugar industry has survived despite the fact that sugarcane prices have risen by 3.9%, compounded year-on-year, more than twice the increase in sugar prices (1.7%) between January 2000 and December 2008; in certain seasons, the absolute price of cane has been pegged at levels higher than that of sugar extracted therefrom. Obviously, taking sugar alone into account, the welfare enhancement of the farmers has been brought about at the expense of the sugar mills' ability to meet capital service obligations. Or, mathematically,

$$
\begin{aligned}
i_t^* G(K) &\equiv [(p_t^*) \times Q - m \times n \times q \times (p_t) - c(K)Q] - F(K) < I^{**} G(K); \\
&\text{while } \pi_k(p_t) \geq \pi_k^*(p_t)
\end{aligned}
$$

Enhanced revenues from power generation and distillery operations have helped nudge $i*G(K)$ beyond $i**G(K)$, a situation often referred to as 'razor-and-blade pricing', where the profitability of supplementary sales enable the producer to offer primary products at discounted prices.

$$i_t^* G_2(K) \equiv [(p_t^*) \times Q + e \times \tau_t + S_t] - [m \times n \times q \times (p_t) + c_2(K)Q] - F_2(K) > I^{**}G_2(K);$$

where variable and fixed costs associated with the power plant and the distillery are included.

Attributed to King Camp Gillette, the razor-and-blade concept involves distributing the base hardware (viz., inkjet printers) or service at near breakeven prices or even at a deep discount, so as to create sustained demand for expensive add on components, software or other consumables (viz., ink cartridges) required to put such hardware to use [5]. Complementary product pricing is used by firms that enjoy high profitability on supplementary sales and value is captured by offering a stripped down version of products (i) to appeal to more price sensitive segments or (ii) to leverage new distribution channels. This strategy is successful when used for a narrow, clearly defined segment in high growth markets, and more importantly, *where price changes are difficult to detect* [30].

The sugar–energy complex is compelled to supply sugar at a discount while, in effect, power generation and distillery operations are required to subsidize sugar production. This strategy is sustainable if, and only if, the producer has control on product pricing and can price the supplementary product(s) substantially above the marginal costs of production. To achieve this, the supplementary product, needs to be non-replicable and non-substitutable, failing which the competition would be willing and able to supply the same at or near marginal cost.

For ethanol to be used as a fuel, its price would be determined by the retail prices of gasoline (petrol), discounted by the relative calorific value. The sugar millers therefore, are faced with a ceiling on ethanol prices. The price of molasses supplied as an intermediate product is in-turn determined by the price of the alcoholic beverages it goes into and by the relative bargaining power of the brewers. Additionally, cane producing states are disinclined to dismantle regulatory controls on alcohol and molasses, which are important sources of revenue [2].

The marginal cost of power generated from other sources viz., large centralized power plants, is generally lower[10] (with CERs compensating for environmental externalities). The tariff offered on power from sugar cogeneration plants varies across states. While some states viz., Kerala and Tamil Nadu offer lower base tariffs and 5% annual tariff escalation for the first few years, others offer higher base tariffs and low or zero escalation. Overall, tariffs from renewable energy projects are computed working backwards from targeted (19% for the first 10 years) normative pre-tax return on equity [11] albeit basing on different

---

[10] Reliance Power Limited has received the concession to develop the 4000 MW, Sasan power project at a tariff of INR 1.19 per kW h [38].

assumptions relating to capital expenditure and other variable parameters of the median plant. Worse, the political establishment could easily transfer the surplus returns to the farmers through the administered price of cane, thereby rendering power and eventually ethanol unviable, along the lines of sugar. For example, cane prices in Uttar Pradesh have been escalated by about 8% compounded year-on-year over the past five years, while the annual tariff escalation paid on power is a more modest INR 0.04/kW h or about 1.4% over the base tariff of INR 2.86.

Clearly, sugar mills are not totally autonomous as far as pricing of ethanol and power is concerned, and substitutes for each could be supplied at marginal costs. Mills cannot price them at a premium to subsidize sugar production. The razor-and-blade pricing thrust upon the sugar industry is therefore not the best fitting strategy.

"This is a country that claims to have beaten Brazil and become the world's largest producer [of sugar]. The government gives sugar to its privileged ration card holders [slated to be poorer sections of society] cheap, but it does not want to spend on the subsidy. So it forces sugar producers to finance it by buying off their sugar at a low price. To stop them from going bankrupt because of this robbery, it lets them sell the rest of their sugar in the open market, and ensures that market prices are high enough to make it worth their while. It makes them hold stocks, but manages the stocks itself—so badly that it has either too much or too little. If a new trade round makes it abandon some of its stupidities, let it be soon!"

Editorial Opinion, Businessworld, 21 September, 2009, p 66.

3.7 Analysis and Discussion

It would be unrealistic to expect a transition to a *laissez-faire* market for cane and sugar, impacting the fortunes of millions of farmers and consumers. Yet, exogenously imposed controls would need to give way to self-regulation to ensure sustainability of the sector. However, such 'depoliticization' of the sugar industry, *inter alia* allowing the markets to determine prices, time the release of sugar, and to decide on exports might not come about in the medium-term.

Given the constraints facing the sugar mills, and with the objective of maximizing collective welfare as in Eq. 3.2, the sugar mills clearly need to focus on the product that is non-replicable and non-substitutable, which in this case happens to be sugar. Ethanol and power could help defray the fixed costs over larger revenue volumes but cannot sustain business operations in the medium-term.

From the above grasp of the interaction between the mill and the farmers located in the catchment area, we now estimate the subgame perfect equilibrium. Clearly, by entering sugarcane cultivation in response to the higher prices on offer (C_m, C_f), the cane farmers create a glut, induce a default by the mill (C_f, D_m), and eventually exit cane cultivation (D_f, C_m), only to return subsequently. Sugar prices have witnessed relatively wild swings as shown in Fig. 3.2. Starting at 146.7 (1993–1994 = 100) for the week of the 9th of September 2000, prices have progressively dropped to 114.1 for the week of the 1st of March, 2003: a 28.6%

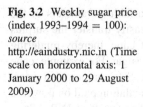

Fig. 3.2 Weekly sugar price (index 1993–1994 = 100): *source* http://eaindustry.nic.in (Time scale on horizontal axis: 1 January 2000 to 29 August 2009)

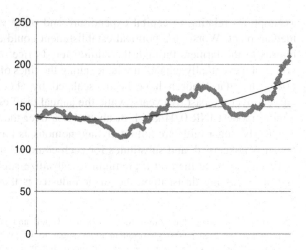

decline over 32 months. Similarly, commencing at a peak of 175.2 for the week of the 18th of February 2006, prices have declined to 140 for the week of 4th August, 2007: a 24% drop over approximately 16 months. Apparently, there has been a glut during the sugar years 2000–2001, 2001–2002. Likewise, there has been excess production of cane/sugar during 2006–2007.

Since individual mills have no control on product pricing or on inventory management under the present regulatory framework, the mills should try and actively stabilize the quantum of cane grown and supplied. This goes hand-in-hand with respecting catchment boundaries and preventing an attritional competition for the available cane. Stabilizing cane output would serve to enhance collective welfare and prevent default by the mills and retaliatory, crop-switching by the farmers.

Given the constraints of the regulatory framework, mills are required to abide by the support- price floors; while mills cannot make cane cultivation *unattractive*, they could work towards making alternative corps *more, or at least equally, attractive*. By rendering the farmer indifferent between growing cane and substitute crops, it is possible to prevent the glut, the default, the defection and the eventual reentry. In summary, ensuring that the acreage under cane remains stable, and hence ensuring that the output of cane is predictable, could help smoothen the sinusoidal sugar price curve to resemble the polynomial 'trend line' in Fig. 3.2. Bridging the valleys, at the cost of losing the peaks should yield positive net benefits for the mill and hence for the farmers, thereby enhancing overall welfare. The unilateral transfers made in the Stage-3 sub-game could be made voluntarily in the Stage-2 sub-game but in order to regulate the influx of farmers switching to cane cultivation.

If 'n' farmers 'cooperate' in Stage-1, production, sugar prices and cash-flow patterns can be stabilized by having 'n' farmers cultivate cane, season after season. The transfers to (N–n) farmers located within the catchment area are made to render alternate crops as attractive as sugarcane. The difference between the returns on cane and the alternate crops viz., paddy, "terms-of-trade" is made up so as to ensure no welfare-loss to the farmers. In doing so, the mill manages to ensure consistency in sugar prices, prevents cash-flow droughts due to low prices on the one hand, and saves on inventory management costs from accumulated output on the other. Securing the cooperation of the '$N-n$' farmers—through their *not* growing sugar cane is vital to the sustainability of the sugar mill. Respecting the exclusivity of the catchment areas and 'cooperating' with fellow mills by not disrupting such transfer arrangements would ensure sustainability of the industry as a whole. In seasons where the nearest substitute corps are more attractive than sugarcane, the mills would have to make the transfers to the 'n' farmers, to in-turn, align their returns to the '$N-n$' non-cane farmers.

Mathematically, the returns for farmers growing other crops are to be equilibrated with returns from growing cane:

$$\pi_k(p_t)_{\text{cane}} = [\pi_k(p_t)_{\text{other}} + h^*] \geq \pi_k^*(p_t); \quad \text{where } H^* = \sum h^*.$$

Subject to:

$$H^* \leq [(p_t^*)_{\text{stabilized}} \times Q] - [(p_t^*)_{\text{volatile}} \times Q] + [PV(p_t - p_{SAP})]$$
$$+ [c_2(K)_{\text{volatile}} Q - c_2(K)_{\text{stabilized}} Q]$$

where PV(.) is the present value of the premium paid over the state advised price.

This proposal is not dissimilar to the production-limiting programs, categorized as "blue-box payments" under the WTO's agreement on agriculture (AoA) and yet merely seeks to maintain stability in acreage under a given crop while helping agriculturists switch crops. Compliance by individual farmers could be assured and monitored by providing technical expertise, loans, farm inputs etc.

This proposal calls for a reinforcement of the reciprocal monopoly between the cane farmers and the sugar mills, and hence for putting an end to cane poaching from the recognized catchment areas of fellow mills.

3.8 Application of the Model

The model so developed has been applied to build a counter-factual scenario, employing published annual financial statements from one of India's largest sugar companies, operating in the northern Indian state of Uttar Pradesh. Sugar prices are projected to evolve more gradually but monotonically, rather than following the sinusoidal pattern observed in reality.

The *Hotelling rule* predicts that unit prices of exhaustible resources (less marginal costs of extraction) tend to rise over time at a rate equal to the return on comparable capital assets [19]. The *Hotelling valuation principle* (HVP) is based on the assumption that in equilibrium, net prices rise at the rate of interest over time and that the rate of interest is equal to the rate of discount. Hence, Miller and

Table 3.5 Rates of sugar price evolution: actual versus simulated

Year	Sep 08(12)	Sep 07(12)	Sep 06(12)	Sep 05(12)	Sep 03(12)	Sep 04(12)	Sep 02(18)
Sugar price realized (INR/t)	16,333.47	14,468.09	18,442.34	17,515.82	12,643.59	15,000.78	14,149.30
Rate of growth (%) (actual)	12.89	−21.55	5.29	16.77	−10.64	18.64	
Growth at risk free rate[#] (%)	6.000	6.000	5.500	4.833	5.125	4.500	

[#]The Reserve Bank of India's (annualized) overnight rates paid to banks—averaged for October (t)–September(t + 1)

Table 3.6 Rates of sugar price evolution: actual versus simulated

Year	Sep 08(12)	Sep 07(12)	Sep 06(12)	Sep 05(12)	Sep 03(12)	Sep 04(12)	Sep 02(18)
Average sugarcane price (INR/t)	1,395.59	1009.56	1,285.23	1,170.08	1,091.29	921.71	1,003.64
SAP[#] for sugarcane (INR/t)	1,250	1,100	1,150	1,120	950	950	950

[#]Uttar Pradesh State advised support price for sugarcane

Upton [28] demonstrate that the per-unit value of exhaustible reserves in the ground is the same as the per unit value when extracted, less extraction costs. This leads to the conclusion that profit-maximizing producers exploiting non-renewable resources should be indifferent between current and future production and that price evolves at the applicable interest (or discount) rate to prevent arbitraging across time. Through their study of an extended sample, Bell et al. [4] confirm the Hotelling theory and the results obtained by Miller and Upton. Though sugarcane is *strictly* not a non-renewable resource, a finite, clearly-assessed quantity is available for harvest each season and hence construed to be "non-renewable" within the limits of each season. The most conservative risk-free rate—the rate at which the Reserve Bank of India borrows from commercial banks (the reverse-repo rate)—is employed to demonstrate the model. Table 3.5 lists the risk-free rate at which the counter-factual sugar prices are projected to rise.

Since the company is slated to have complete control on the quantum of sugarcane grown within its command area and that it faces no competition from fellow mills (or from the informal sector), the company is not required to pay a premium over the state advised price (SAP) for the respective year (Table 3.6). The other sources of revenue are relatively minor (and their unit prices are fixed as in the case of power, or limited by substitutes as in the case of ethanol) and are retained as in the original statements of account. Likewise, the expenditure pattern has been mapped in proportion to the simulated expenditure on sugarcane.

Table 3.7 lists the results from simulating the revenues and expenditure for the seven-year horizon. A part of the company's debt is retired during the 18 month period ending September 2002 (mentioned at actual), while additional debt is taken on during the subsequent years (not added to cash-flow available for

Table 3.7 Simulated counter-factual scenario

Year	Sep 08(12)	Sep 07(12)	Sep 06(12)	Sep 05(12)	Sep 03(12)	Sep 04(12)	Sep 02(18)
Reported profit after tax	297,996,005	462,665,477	1,063,377,641	717,390,390	833,348,726	511,521,611	167,398,058
Extraordinary items	73,445,783	2,288,134	−19,237,876	15,123,760	2,002,214	−10,616,186	−44,961,362
Adjusted net profit	224,550,221	460,377,342	1,082,615,517	702,266,629	831,346,511	522,137,798	212,359,420
Add back depreciation	1,676,892,635	1,600,386,563	647,734,816	335,881,491	166,183,820	139,247,264	174,450,087
Deduct debt repayment							−1,273,200,000
Cash available for distribution	1,901,442,856	2,060,763,905	1,730,350,333	1,038,148,120	997,530,331	661,385,061	(886,390,492)

Table 3.8 Reported Actual Figures for the Company

Year	Sep 08(12)	Sep 07(12)	Sep 06(12)	Sep 05(12)	Sep 03(12)	Sep 04(12)	Sep 02(18)
Adjusted net profit	−558,800,000	454,400,000	1,929,800,000	1,388,100,000	607,900,000	293,800,000	151,100,000
Add back depreciation	1,872,200,000	1,468,800,000	723,900,000	350,900,000	190,900,000	135,100,000	184,300,000
Deduct debt repayment							−1,273,200,000
Cash available for distribution	1,313,400,000	1,923,200,000	2,653,700,000	1,739,000,000	798,800,000	428,900,000	(937,800,000)

appropriation). Table 3.8 reproduces the respective year-end figures reported by the company.

For *average* annual increases in sugar price in excess of 5.33% (real rate) the NPV[11] (as of October 2002), of the simulated cash-flow stream is greater than that of the reported actual cash-flow stream: a part of this surplus could be used to adjust the relative terms-of-trade, to in-turn bring about such a gradual price evolution. In reality, the risk-free rate pertaining to a 12 month bank deposit (as opposed to the relatively risk-free benchmark rate applied by the country's central bank) would be more appropriate to benchmark annual sugar price increments.

3.9 Conclusions

This chapter begins by establishing the pricing trail from the raw materials to the finished product and vice versa. It is observed that in recent years, consequent to the diversification strategies adopted by sugar mills, sugarcane prices are linked to electricity prices, rather than to sugar production which, over time, is deemed to be a by-product. This chapter asserts that power generation and ethanol production cannot resurrect an otherwise unviable sugar operation. Industry and product characteristics do not permit the use of razor-and-blade (complementary product) pricing since the sugar mills are neither exclusive nor least-cost producers of power, while power and ethanol do not supplement sugar. While a diversified product offering could garner additional revenues, the industry's survival—and by extension, power generation and ethanol production—would require a healthy sugar segment.

An iterated prisoner's dilemma game is employed to represent the interaction between the sugar mills and the farmers. This is treated as a game with asymmetric payoffs since the mills lack control on pricing for cane, while the farmers have "outside options", i.e., the option to switch crops. The mill is pictured as the "disadvantaged" player who makes the first move by investing in the specific-purpose fixed-asset. Hence, the mill tends to be more "inertial" in decision making while the farmer makes the "cooperate/defect" decision each season.

This chapter demonstrates that the influx of a large number of farmers (their decision to "cooperate", by definition) leads to lower sugar prices, to constrained cash-flows for the mill, and an eventual default on the payments owed to the sugarcane farmers. In the next stage of the game, the farmers play tit-for-tat and "defect" by switching to other crops. In reality, the farmers' decision to "cooperate" and thereby to produce surplus cane leads to unfavorable outcomes for all concerned, and hence to diminished cumulative welfare. Further, in subsequent stages of the game, mills end-up bidding higher prices for the available cane grown

---

[11] Discounted at 11%, the long-period average-(real)-equity-returns earned on the Bombay Stock Exchange benchmark index (BSE Sensex) [26].

within partially overlapping command areas. This is represented as a unilateral transfer from the mill to the farmer in the form of a scarcity rent.

Believing that such "sugar cycles", i.e., periods of surplus production followed by several seasons of diminished output, are not inevitable, a model is developed to help sugar prices evolve gradually and monotonically at the rate of the applicable risk-free interest rate, in accordance with the Hotelling Rule. Given that companies can maintain the reciprocal monopoly with the sugarcane growers and that they would respect the earmarked command areas, it should be possible for the sugar mills to ensure that *just* the right quantity of cane is grown. In order to maintain such precision, and as a direct consequence, the mills could adjust the terms-of-trade of other crops, viz., paddy, wheat etc. by making them less or more attractive relative to cane, as required.

The model so developed is applied to one of India's largest and fastest growing (measured by installed cane crushing capacity) sugar companies operating in the northern Indian state of Uttar Pradesh. With the help of the simulated counterfactual scenario, this chapter demonstrates that an increase in the market price of sugar at conservative rates still leaves a surplus (relative to the actual reported figures) for the mill to deploy towards adjusting the relative terms-of-trade.

This chapter therefore recommends strengthening the exclusivity of the command areas reserved for mills, and a more intensive engagement with the farmers therein; and non-compete agreements among mills enforced by the respective statutory authorities. Sugar mills could project a gradual increase in the market price of sugar, in contrast to the wild swings witnessed in the past decade. Likewise, the mills pay the state advised price (SAP) for cane but are *not* required to pay a premium, owing to secured reciprocal monopoly with the farmers. The savings could then be shared with the farmers (a) to ensure that just the right number grow sugarcane and more importantly (b) to prevent an influx of farmers into sugarcane cultivation, in response to revised support prices. This is similar in spirit to the "blue box" subsidies listed in the Agreement on Agriculture, negotiated at the World Trade Organization (WTO). In this case, crop limitation is achieved by encouraging cultivation of alternative crops and reciprocally, when sugarcane is deemed unattractive, by compensating the farmers for the deficit in returns. A financially viable and solvent sugar operation is a prerequisite for the mill to successfully diversify into power generation from bagasse and ethanol production, and can be achieved through the suggested measures.

Even as the statistics quoted and the analyses conducted pertain, in detail, to the sugar industry, the central concept of scarcity rents (including those surpluses emanating from preferential tariffs and fiscal incentives) migrating backwards, is equally applicable to the solar photovoltaic industry (profits migrating backwards to silicon wafer producers and to production-line equipment makers), to wind turbine manufacturers and owners of wind-resource sites in the wind-energy sector, to the manufacturers of hydro-mechanical equipment and owners of resource-rich sites in the case of small-hydro power plants etc. A transparent auction of the sites or other scarce resources would ensure that the scarcity rents are effectively captured and distributed while ensuring optimal utilization of the scarce resource in question.

References

1. Ahmed F (2009) Sugar: bitter sweet. Businessworld, p 13. Accessed 7 Sept 2009
2. Allirajan M (2008) Sugar: sweet options. Businessworld. Accessed 18 Jan 2008
3. Banerjee A, Mookherjee D, Munshi K, Ray D (2001) Inequality, control rights, and rent seeking: sugar cooperatives in Maharashtra. J Political Econ 109(1):138–190
4. Bell FW, Lacombe DJ, Ryan MP, May DL (2000) An empirical re-examination of the hotelling valuation principle. J Econ Res 5:1–15
5. Bulkeley WM (2007) Kodak's strategy for first printer—cheaper cartridges. The Wall Street J. Accessed 6 Feb 2007
6. Charness G, Frechette GR, Qin C-Z (2007) Endogenous transfers in the prisoner's dilemma game: an experimental test of cooperation and coordination. Games Econ Behav 60(2):287–306. 2006
7. Dutta AP (2009) Indian sugar mills generate as much 'Green' energy as windmills, and at half the cost. Down Earth 18(2), Accessed 27 May 2009
8. Fujiwara-Greve T, Yasuda Y (2009) Cooperation in repeated prisoner's dilemma with outside options. 20 June 2009. http://ssrn.com/abstract=1092359
9. Gemmill G (1980) Form of function, taste and demand for sugar in seventy-three nations. Eur Econ Rev 13(2):189–205
10. Ghosh S (2002) Electricity consumption and economic growth in India. Energy Policy 30(2):125–129
11. Gipe P (2009) India's 1.1 billion move to feed-in tariffs. RenewableEnergyWorld.com. Accessed 6 Oct 2009
12. Girvan NP (1987) Transnational corporations and non-fuel primary commodities in developing countries. World Dev 15(5):713–740
13. Godbole M (2000) Co-operative sugar factories in Maharashtra—case for a fresh look. Econ Polit Weekly 35(6):420–424
14. Granger CWJ (1969) Investigating causal relations by econometric methods and cross-spectral methods. Econometrica 34:424–438
15. Harris S (1987) Current issues in the world sugar economy. Food Policy 12(2):127–145
16. Haya B, Kirpekar S, Ranganathan M (2005) A sweet choice for power? evaluating climate change financing for sugar mill cogeneration in India. University of California at Berkeley, Breslauer symposium, Paper 12. http://repositories.cdlib.org/cgi/viewcontent.cgi?article=1019& context=ucias
17. Hearne RR, Easter WK (1997) The economic and financial gains from water markets in Chile. Agric Econ 15(3):187–199
18. Hoffmann R, Ging LC, Ramasamy B, Yeung M (2005) FDI and pollution: a granger causality test using panel data. J Int Dev 17(3):311–317
19. Hotelling H (1931) The economics of exhaustible resources. J Polit Econ 39:137–175
20. Hughes H, Singh S (1978) Economic rent: incidence in selected metals and minerals. Resour Policy 4(2):135–145
21. Jishnu L, Srivastava S (2005) Sugar: cane wars; Sugar is suddenly hot. And so is the fight for capacity and leadership. Businessworld. Accessed 29 Aug 2005
22. Kilima FTM (2006) Are price changes in the world market transmitted to markets in less developed countries? A case study of sugar, cotton, wheat, and rice in Tanzania, IIIS Discussion Paper No.160, June 2006
23. Krivonos E, Olarreaga M (2006) Sugar prices, labour income and poverty in Brazil. World Bank Policy Research Working Paper 3874, April 2006

24. Liu LC, Hsu C Ed, Younis MZ (2008) The association between government expenditure and economic growth: granger causality test of US Data, 1947–2002. J Public Budgeting, Accounting and Financial Management, 20(4):439–452. http://works.bepress.com/cgi/viewcontent.cgi?article=1020&context=edhsu
25. Mbohwa C, Fukuda S (2003) Electricity from bagasse in Zimbabwe. Biomass and Bioenergy 25(2):197–207
26. Mehra R (2006) The equity premium in India. National Bureau of Economic Research, NBER Working Paper 12434, August 2006
27. Menon S (2009) Merchant power: to cap or not to cap. Businessworld p 22. Accessed 14 Sept 2009
28. Miller MH, Upton CW (1985) A test of the hotelling valuation principle. J Polit Econ 93(1):1–25
29. MNRE (2008) Annual report of the ministry of new and renewable energy, Government of India, 2007–2008, p 47
30. Noble PM, Gruca TS (1999) Industrial pricing: theory and managerial practice. Mark Sci 18(3):435–454
31. Nopany C (2008) Insider view: nothing sweet about it, Businessworld, 2008. http://www.businessworld.in/index.php/Nothing-Sweet-About-It.html
32. Pagoulatos E, Sorensen R (1986) What determines the elasticity of industry demand? Int J Ind Org 4(3):237–250
33. Pirard R, Irland LC (2007) Missing links between timber scarcity and industrial overcapacity: lessons from the Indonesian pulp and paper expansion. For Policy Econ 9(8):1056–1070
34. Purohit P, Michaelowa A (2007) CDM Potential of Bagasse Cogeneration in India. Energy Policy 35(10):4779–4798
35. Rowse JG, Copithorne LW (1982) Natural resource programming models and scarcity rents. Resour Energy 4(1):59–85
36. Smith RP, Sola M, Spagnolo F (2000) The prisoners' dilemma and regime-switching in the Greek–Turkish arms race. J Peace Res 37:737–750
37. Smouse SM, Staats GE, Rao SN, Goldman R, Hess D (1998) Promotion of biomass cogeneration with power export in the Indian sugar industry. Fuel Process Technol 54(1):227–247
38. Subramani MR (2006) Mills wrangle for sugarcane in UP. The Hindu—BusinessLine, 17th Jan 2006. http://www.thehindubusinessline.com/2006/01/17/stories/2006011701651200.htm
39. Subramaniam K (2009) The power of shortage. Businessworld, pp 38–39. Accessed 28 Sept 2009
40. Sunnevag KJ (2000) Designing auctions for offshore petroleum lease allocation. Resour Policy 26(1):3–16

Chapter 4
Revealed Preferences and the Power of Substitutes

Socialism proposes no adequate substitute for the motive of enlightened selfishness that today is at the basis of all human labor and effort, enterprise and new activity
William Howard Taft
27th President, and later 10th Chief Justice of the USA.

Liquid biofuels—ethanol and biodiesel—are widely recognized, technically feasible alternatives to fossil fuels. Even as the jury is out to determine the environmental footprint of biofuels, the surrounding frenzy has often led to the announcement of unsustainable support prices for feedstock and unviable procurement prices for the finished product. Taking on from the discussion on scarcity in the previous chapter, the following chapter makes a detailed assessment of incentive structures facing agriculturists, refiners and the consumers. Data from the Indian market are employed to illustrate the power of substitutes.

"The ease of manufacturing biodiesel from vegetable oils and animal fats has made it one of the most promising, near-term alternatives to fossil fuels" [15]. Liquid fuels made from biomass—biodiesel and ethanol—could be substituted for petroleum–diesel and gasoline respectively. This comes as a blessing to oil importing developing nations, who now actively consider refining vegetable oil into biodiesel and using it to meet domestic consumption or for export, and in the process, hope to improve their trade balances and to conserve on foreign currency reserves. Ethanol is slated to be "greener, cheaper, more secure than gasoline" and a switch is unlikely to cost the consumer, automakers or the government anything [23]. The rapid rise in crude oil prices and the geo-political uncertainty associated with ensuring uninterrupted supplies have compelled urgent action on this front.

This chapter is largely based on, Srinivasan S (April 2009) The food v. fuel debate: a nuanced view of incentive structures, Renew Energy, The Official Journal of the World Renewable Energy Network, Elsevier, Vol. 34(4).

Aeck [1] cites statistics published by the International Energy Agency (IEA) to show that Ethanol production nearly doubled between 2000 and 2004, while bio-diesel production rose from a mere 11 million liters in 1991 to 1.77 billion liters in 2003. Countries as geographically dispersed as Thailand, Uruguay and Ghana could potentially take the lead in biodiesel production in the estimated 51 billion liter industry, approximately displacing 4–5% of the current global petroleum–diesel consumption. Brazilian sugarcane ethanol, which is often showcased as an ideal to be emulated, is cheap to produce has an encouraging energy pay-back and hence, meets 40% of the country's fuel needs [19]. On the other side of the globe, owing to weather conditions perennially favoring the growth of the oil palm, Malaysia today, is the largest producer and exporter of palm oil, accounting for over 5% to the country's GDP. It is also claimed that palm oil requires less inputs viz., fertilizers, pesticides, herbicides and energy, per ton of oil, compared to alternatives such as soybean, rapeseed and sunflower [32].

Corn-based ethanol, which has until now been central to US biofuel policy, may not be the most suitable alternative to fossil fuels. So called "second-generation" cellulosic ethanol is made efficiently using powerful catalysts and enzymes and could use waste products such as straw, corn stalks or agricultural debris [9]; this despite the fact that plant biomass has evolved over millions of years, developing resistance to biological enzymes [25]. It is estimated that the energy efficiency and pollution characteristic of cellulosic ethanol, favored by Canada, could be far superior to that of corn ethanol encouraged by the United States. While Europe has extracted first generation ethanol largely from wheat and sugar beet [7], maize farmers in South Africa hope to channel surplus output into ethanol [27]. Research on "third-generation" biofuels is already underway in different parts of the world.

4.1 The Environmental Footprint of Biofuels

Activists and to a large extent, policy makers, seem to be ambivalent on the environmental viability of biofuels. Ethanol produced from "cellulosic" feedstock, agro-residues and fast growing hays like switchgrass, etc., is slated to yield a positive energy balance. It is also argued that the biomass merely returns the recently sequestered carbon dioxide to the atmosphere and hence such combustion is carbon-neutral. Deep-rooted perennials are also found to retard soil erosion [17]. Others contend that a 50 million gallon per annum ethanol factory consumes 500 gallons of water a minute for its boiling and cooling processes, a part of which is lost through evaporation and waste discharge [12]. Further, the heat generated from the combustion of a liter of ethanol is about two-thirds that from a liter of petrol, thus necessitating a larger quantity of fuel to derive the same energy output. Further, ethanol itself requires to be blended with petrol to avoid corrosion and rapid deterioration to seals etc. [10]. Consequently, Charles et al. [6] believe that the heightened reliance on biofuels may potentially "inhibit the development and maturation of longer-term alternatives" that could mitigate fossil-fuel dependence.

Oil seed bearing *Jatropha curcas Linnaeus*, ("Jatropha") is seen as an effective solution to combat the greenhouse effect, help mitigate soil erosion, provide rural employment and higher incomes and as a source of energy. Yet, [28] call for a proactive and simultaneously cautious approach to large-scale biofuel cultivation and a rigorous debate on the desirability of biofuels as a major energy source. *Ceteris paribus*, "to be a viable alternative, a biofuel should provide a net energy gain, have environmental benefits, be economically competitive, and be producible in large quantities without reducing food supplies". Palm biodiesel from parts of Indonesia is considered unsustainable because of the displacement of large tracts of virgin rain forest and peat lands, while soybean cultivation has been replacing the rich biodiversity of the Amazon in South America [30]. Blanco and Azqueta [5] conclude that the environmental performance of biofuels is determined almost entirely by the agricultural practices adopted during cultivation.

It is now confirmed that clearing of land in favor of biofuel crops and the consequent loss of forests, peatlands and grasslands would actually aggravate global warming and climate change [16]. The controversy with biofuels is not restricted to their environmental footprint. It is further argued that, biofuels cannot replace much petroleum without impacting food supplies: dedicating all US corn and soybean production to biofuels would meet only 12% of gasoline demand and 6% of diesel demand [20], and would aggravate the uncertainty of supplies, compared to imported oil [8]. Globally, seven crops, namely, wheat, rice, corn, sorghum, sugarcane, cassava and sugarbeet, collectively account for 42% of the cropland, and diverting the entire supply of these seven crops to bioenergy production would satisfy just over half of the global gasoline consumption [26].

4.2 The Food v. Fuel Debate

The genesis of the food v. fuel debate is the rapid rise in coarse grain production—corn, barley, sorghum and other grain used as fodder—compared to primary food grains such as wheat and rice, and the quantum leap in converting these coarse grains to biofuels and their use as livestock feed. Global food grain production has tripled since 1961, and after adjusting for the doubling of the human population over the same period, the per capita production stands at 350 kg in 2007. Human and livestock consumption at 48 and 35% respectively, leave approximately 17% of the world's grain for conversion to ethanol and other fuels [31].

Even as the competitive dynamic among food, fossil fuels and bio-fuels is a singular blend of politics and economics, the rapid expansion of biofuel production from maize, sugarcane, oil-seed and from conversion of edible oils has raised serious concerns on preserving the food security of the planet. Such substitution is either driven by diversion of agricultural produce or by altering cropping patterns themselves. Rajagopal et al. [26] believe that a large-scale switch to "energy plantation" is likely to "induce structural change in agriculture and change the sources, levels, and variability of farm incomes."

4.3   Indian Biofuel Scenario: A Case in Point

A large oil importer, India has consciously sought indigenous alternatives in recent years, to fuel its rapid economic growth. Sugarcane is the most important crop in India after rice and wheat and the country is second only to Brazil in annual sugarcane output [24]. Ethanol is a by-product from molasses in Indian sugar mills and the residual after providing for industrial use and consumption alcohol is available for blending with fuels. Increasing the acreage under sugarcane to produce ethanol directly from sugarcane juice as compared to the sugar—molasses—ethanol route is viewed as being profitable for farmers and sugar mills, while also simultaneously tackling social concerns in certain parts of the country [2]. Others favor the use of sweet sorghum for its significantly higher annual yields, short growth cycle and lower input costs and significantly lower water requirement, as compared to sugarcane [21, 29]. India also produces an estimated 800 mt of agro-residues and converting a small fraction of this into ethanol could supplant current petrol requirements.

The possibility of converting Jatropha (locally known as *ratanjyot*) and Pongamia Pinnata (*"karanj"*) has, over time, captured the imagination of researchers, NGOs and policy makers alike, and has helped bring over 11 million hectares under jatropha cultivation. The National Rural Employment Guarantee Scheme[1] (NREGS) has sought to create employment for unskilled agricultural labor through extensive jatropha and pongamia cultivation. The state of Chhattisgarh, which has take a lead in propagating biodiesel has announced a minimum support price of INR 6.5 kg^{-1} ($\sim$US 16 cents) of jatropha seed, while the market prices touch INR 30 or more a kilogram ($\sim$US 79 cents) [7, 18], owing to lower than anticipated harvests. With an oil yield of 30–35% of seed crushed, three kilograms of seed are required to produce a liter of biodiesel; lower than projected yields and higher prices tend to render biodiesel unviable for distillers and retailers.

India hopes to convert "marginal lands" and to employ surplus agricultural labor to produce the non-edible oil required for biodiesel production. However, given the high cost of inputs and scarcity of feedstock availability, Indian entrepreneurs prefer to import cheaper palm oil from South-East Asia instead [22]. Additionally, Asher [4] points out that land categorized as common property resource provides for livestock rearing, fuel wood, fodder, non-timber forest produce, and hence form an integral part of the rural economy, even as the government classifies such lands as "wasteland". In other instances, "wastelands" are brought under cultivation by farmers who are yet to receive legal title for the same.

---

[1] www.nrega.nic.in.

4.4 An Analysis of Incentive Structures

Rajagopal et al. [26], model the change in the price of gasoline "dP" as a function of the change in the supply of biofuels "dS^E"-ethanol, biodiesel etc.

$$dP = \frac{1}{(\xi_D - \xi_S)} \frac{dS^E}{Q} P$$

where $\xi_{D,S}$ are the elasticities of demand and supply for fuel, and P,Q are the price and quantity of fuel, respectively.

The model makes significant implicit assumptions:

- The biofuel production decision is independent of the prevailing consumer price of gasoline/diesel and that its price is exogenously determined.
- Biofuel production is assumed to push gasoline/diesel prices lower, in its role as a potential substitute and conversely, the price of gasoline is slated to rise as biofuel production is scaled down.

4.4.1 End-User Price

In a free market, the rational, utility maximizing consumer would choose a transportation fuel between gasoline (diesel) and its substitute, ethanol (biodiesel) so as to arrive at an indifference level, wherein *the cost of traversing a mile is the same irrespective of the fuel used.* By this measure, the pricing of substitutes—biofuels—would be determined by their calorific value in comparison to gasoline/diesel. If, for instance, biodiesel were to yield the same mileage as petro-diesel, combined with superior environmental characteristics and lower engine-wear, its consumer price should be the same, if not marginally higher, than the market price for petro-diesel. The retail price of biofuel used as a dope or a substitute, therefore is endogenously determined. The Indian Planning Commission, for instance, has estimated that biodiesel, basing on production costs, could be procured at INR 25 a liter [14], a price substantially lower than the retail price of petro-diesel, and in the absence of credible price fences and an effective enforcement mechanism, creating an incentive for the refiners to sell directly to end users.

The Federal Ministry of Petroleum and Natural Gas in India has determined the procurement price for ethanol based on production costs ("cost plus" approach) at INR 18.75 per liter [22]. This when adjusted for the lower calorific value in comparison to gasoline—60%—amounts to an equivalent value of INR 31.25 per liter of gasoline. The post-tax end-user price of ethanol has, *necessarily*, to match the retail price of gasoline: a lower price would create an opportunity for ethanol brewers to blend fuels by circumventing the government tendered procurement; a higher price, while encouraging larger quantity of output, would discourage procurement by the oil marketing companies or by end users.

4.5 Biofuel as a Substitute

As discussed above, pushing biofuel cropping to the limits of agriculture would merely displace an insignificant proportion of fossil fuel consumption, while leaving the planet with no food. In effect, the first generation biofuels, as known to mankind could, at best, help reduce dependence on (imported) oil, on the margin. It is also observed that biofuel refining is economically viable only beyond a threshold price of benchmark crude oil. It is therefore not inconceivable for the oil exporting nations to drop oil prices sufficiently, to discourage investments in refining operations—and push prices higher when the threat has passed. Such reductions in prices of fossil fuels, owing to ethanol and biodiesel production, therefore, are likely to be marginal and short-lived, if at all.

If the bio-fuel price is endogenously determined and the production decision *is* a function to retail prices of fossil fuels, the market is cleared when:

$$S^T(P) = D^F(P)$$

$$S^G(P) + S^B(P) = D^F(P)$$

where $S^T(P)$ is the total fuel supply at price P; $D^F(P)$ is the total fuel demand at price P; $S^G(P)$ is the supply of fossil fuels at price P and $S^B(P)$ is the production function for biofuels at price P.

The first order condition gives us the response to a change in end-user price:

$$S_P^G + S_P^B = D_P^F$$

where S_P^G is the change in supply of fossil fuels with price—dS^G/dP; S_P^B is the change in supply of biofuels with price—dS^B/dP; D_P^F is the change in demand for fuel with price dD^F/dP.

On rearranging the terms:

$$S_P^B = D_P^F - S_P^G$$

Elasticity of fuel demand

$$\zeta_D^F = D_P^F \cdot (P^0/Q^0)$$

Elasticity of fossil fuel supply

$$\zeta_S^G = S_P^G \cdot (P^0/Q^0)$$

where P^0 and Q^0 are initial price and quantity, respectively, determining the equilibrium mix between fossil fuels and bio-fuels.

$$dS^B = Q^0 \cdot (\zeta_D^F - \zeta_S^G) \cdot (dP/P^0)$$

$$dS^B = f(\zeta_D^F - \zeta_S^G)$$

The change in bio-fuel production is a function of the respective elasticities of fuel demand and fossil fuel supply. In most contemporary societies, demand for transportation fuel tends to be relatively inelastic in the short run. An escalation in price encourages an increase in bio-fuel output, which could be compensated for by a reduction in import/supply of fossil fuels. Bio-fuels could thus satisfy a larger proportion of the incremental demand over time.

4.6 Revision of End-User Prices

The prices of petroleum products in India, as in several other developing countries, are caught in a web of cross-subsidies and normative pricing, as refined products such as petrol (gasoline), diesel, kerosene etc., are retailed at prices determined by the federal government. Despite the cosmetic restructuring of the administered price mechanism (APM), over the years, the upward revision in retail prices of distillates to attain import parity, necessitated by rising international crude oil prices, continues to test the government's ability to push through politically unpalatable decisions in the larger interest of the economy. Retail fuel prices, which tend to feed into the inflation in commodity prices faced by the voter–consumer, are therefore driven, more by political considerations than by economic compulsions from increasing input prices. This conflict of interests has been exposed during periods, when benchmark prices of crude oil in the international markets have touched historic highs, well in excess of US$100 per barrel, on occasion.

Employing weekly spot prices published by the US Government's Energy Information Administration,[2] and the corresponding US Dollar—Indian Rupee exchange rates published by the Reserve Bank of India (RBI), the weekly Indian-Rupee-denominated spot prices have been simulated. It is observed that the coefficient of determination (r-squared) for the OLS regression between the simulated spot prices and the index of petrol (gasoline) and diesel prices[3] is 0.875 and 0.850 respectively, at the 95% confidence level. In effect, retail prices of petrol (gasoline) and diesel do not mirror international crude oil prices and end-user prices in the medium term are stable and predictable.

As a direct consequence of the administering retail prices of the distillates, the government has also involuntarily administered the prices of substitutes. *Essentially, fossil fuel distillates determine biofuel prices and not vice versa.* A change in retail prices of the petro-fuels, when announced, would also impact the procurement prices of bio-fuels, indexed for the difference in calorific values.

---

[2] http://tonto.eia.doe.gov/dnav/pet/hist/wtotworldw.htm.

[3] http://eaindustry.nic.in/; Economic Advisor to the Ministry of Commerce and Industry, Government of India.

Bio-fuel prices are expected, therefore, to move in tandem with international benchmark crude prices (with an overlap of about 85% in India), offering an incentive for the biofuel refiners to export their produce to markets where retail petro-distillate prices mirror higher benchmark crude oil prices *in toto*. In this scenario, landed biofuel prices would be determined by the prevailing fossil fuel retail prices in the target markets. Eickhout et al. [13] conclude that environmental and trade agreements must be consistent to ensure adequate coordination and to improve the environment while receiving the benefits of free trade; such coordination would eliminate perverse incentives which could lead to environmental exploitation in poor countries, in order to benefit from higher fuel prices in export markets.

4.7 Refining Margins and Farmers' Incentive Structures

Working backwards from the price of the finished product, and the costs involved in transesterification and related processes, refiners choose to procure feedstock from farmers' (or cooperatives) at a price that yields a competitive return on their investment, which is comparable with the long-run average return on a broad-based stock market index. Additional revenue streams could accrue from the sale of byproducts viz., glycerine, oil cake, and from trading approved carbon emission reduction.[4] Exogenously determined support prices for seed procurement in the case of biodiesel, and sugarcane in the case of ethanol, with a constant threat of upward revision, would make it unattractive for refiners to invest in the business. On the other hand, such inflated support prices encourage farmers to convert to biofuel plantation, at the cost of other food and non-food crops.

The "cut-in" prices that would induce cultivation of biofuel crops, would be determined by the indifference level of the farmers, measured by a benefit-cost metric between biofuel and the alternatives, other factors such as knowledge, credit, fertilizer and irrigation held constant. Assuming that an individual farmer's decision to grow one or the other crop does not influence its market price, the support price announced for biofuel crops and the price for food crops viz., paddy, wheat etc., discovered on the futures markets would help the farmer compute and compare such benefit-cost measures.

$$\frac{\text{Benefit}_{(\text{Food})}}{\text{Cost}_{(\text{Food})}} > = < \frac{\text{Benefit}_{(\text{fuel})}}{\text{Cost}_{(\text{fuel})}}$$

It is, therefore, possible for policy makers to encourage or control the use of finitely available cropland for the cultivation of food crops or alternatively for energy crops, through appropriate price signals. The "cut-out" price that

---

[4] See for instance, cdm.unfccc.int/UserManagement/FileStorage/FS_686206579.

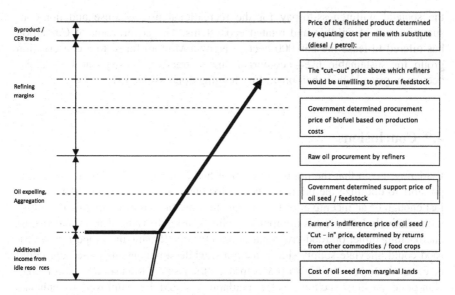

Fig. 4.1 Growth in biofuel feedstock production in tandem with increase in price

discourages expansion of biofuel cultivation is, ironically, an even higher support price for the crop, high enough to discourage refiners from procuring any additional quantities, as shown in Fig. 4.1.

As fossil fuel prices are revised, the real increase in price of bio-fuels cascades down to the refiners, the oil expellers and the feedstock farmers in the proportion of their original stakes in the prices of the finished goods and in keeping with the relative scarcity of their capacities, as discussed in the previous chapter. If support prices on food corps remain constant, higher prices would encourage more farmers to make the switch from food-crops to fuel feedstock. The converse would apply just as well.

4.8 Marginal Lands and Conversion of Forest Land

It is generally perceived that *jatropha* is hardy and grows on marginally fertile lands, requires lower irrigation and application of fertilizers and grows largely unattended. There is, thus, interest in bringing such marginal lands under the plough, while existing cropland is left intact to grow food-crops. However, on conversion, such marginal lands and pastures appreciate in value as agricultural lands and farmers face the same incentive structures as before and make their choices based on the economic incentives.

Loss of naturally occurring biodiversity and spread of monoculture is a major environmental concern. Palm oil farmers in South-East Asia and Soybean farmers in South America face two levels of incentives, immediate revenues from the sale of timber and long-term revenues from fuel crops. Allen and Loomis [3] estimate the societal willingness-to-pay for the recreational use, non-use and non-consumptive use of wildlife and related ecosystems. The government of Cameroon has offered to lease the 830,000 hectare Ngoyla-Mintom forest to a conservation group for sustainable use—ecotourism and extractive forestry—at US$ 2 per hectare, a price higher than prevailing logging concessions [11].

4.9 Conclusions

Evidently, given that the pricing of the finished product itself is less flexible, it is possible to optimize the pricing structure for fuel feedstock to ensure that there is just enough of an incentive for the farmers to grow fuel crops on marginal lands—with relatively lower levels of inputs—while creating the right disincentive against encroaching either on food-crop land or onto forests. Raising the support prices for food could alleviate supply side constraints, while simultaneously the poor could be compensated for the higher prices through the issue of food stamps and the like. Non-price barriers such as water availability could be employed to enhance rotation between food and fuel crops over the agricultural calendar.

Even as the de facto price-ceiling for bio-fuels is discussed extensively in this chapter, similar incentive structures are faced by other renewable energy technologies, as well. For instance, the power tariff from solar photovoltaic systems and wind turbines is benchmarked against the end-user price for grid-supplied electric power, which is often subsidized for many consumer segments. Likewise, the price-effectiveness of solar thermal systems is measured against the alternatives available to heat interior spaces or water, as with LPG-fired or electric geysers. By administering or exogenously determining the price of a given product or service, policy makers inadvertently set price ceilings for potential substitutes, as well. While offering subsidies and other fiscal incentives to enhance the acceptance of the more expensive (albeit environmentally sound) alternative is an immediately available option, the real long-term solution might lie in leveling the playing field and by allowing the markets to determine prices of various alternative product/service options.

References

1. Aeck M (2005) Biofuel use growing rapidly. Vital signs. Worldwatch Institute, Washington
2. Aiyar SA (2005) From robber barons to sugar barons, *Times of India*, Mysore Edition, 4 September 2005
3. Allen BP, Loomis JB (2006) Deriving values for the ecological support function of wildlife: an indirect valuation approach. Ecol Econ 56:49–57
4. Asher M (2006) Don't call it wasteland, *The Times of India*, Mysore Edition, 12 October 2006
5. Blanco MI, Azqueta D (2008) Can the environmental benefits of biomass support agriculture?—The case of cereals for electricity and bioethanol production in Northern Spain. Energy Policy 36:357–366

6. Charles MB, Ryan R, Ryan N, Oloruntoba R (2007) Public policy and biofuels: the way forward? Energy Policy 35(11):5737–5746
7. Dasgupta A (2007) Alternative energy: fuelling change. *Businessworld*, 3 December 2007, pp 88–90
8. Eaves J, Eaves S (2007) Renewable corn-ethanol and energy security. Energy Policy 35(11):5958–5963
9. Economist (2006) The ethanol: life after subsidies, *The Economist*, 9 February 2006
10. Economist (2007) The advanced biofuels: ethanol, schmethanol. *The Economist*, 27 September 2007
11. Economist (2008) The price of conservation: the unkindest cut. *The Economist*, 14 February 2008
12. Economist (2008) The ethanol and water: don't mix. *The Economist*, 28 February 2008
13. Eickhout B, Van Meijl H, Tabeau A, Van Rheenen T (2007) Economic and ecological consequences of four European land use scenarios. Land Use Policy 24(3):562–575
14. Express (2005) The new money plant, *The New Sunday Express*, 4 December 2005
15. Fisher M (2007) Researchers examine world's potential to produce biodiesel, *News*, University of Wisconsin, Madison. http://www.news.wisc.edu/14307. Accessed 17 October 2007
16. Flavin C (2008) Time to move to a second generation of biofuels. Worldwatch Institute, Washington
17. Glasgow N, Hansen L (2005) Setting the record straight on ethanol, *Renewable Energy Access*. http://www.renewableenergyworld.com/rea/news/reinsider/story?id=38601. Accessed 14 November 2005
18. Handique M (2007) Jatropha may secure energy, but planting schemes fail to bear fruit, *liveMint.com*. 21 November 2007
19. Hasan R (2006) The ethanol investment craze, *Renewable Energy Access*. http://www.renewableenergyaccess.com/rea/news/story?id=45231. Accessed 26 June 2006
20. Hill J, Nelson E, Tilman D, Polasky S, Tiffany D (2006) Environmental, economic and energetic costs and benefits of biodiesel and ethanol biofuels. www.pnas.org. Accessed 25 July 2006
21. ICRISAT (2007) Pro-poor biofuels outlook for Asia and Africa: ICRISAT's perspective, Working Paper. www.icrisat.org/Investors/Biofuel.pdf. Accessed March 2007
22. Jishnu L (2006) Missing the gold in the green, *Businessworld*, 14 August 2006, pp 48–54
23. Khosla V (2006) Face value: a healthier addiction, *The Economist*, 23 March 2006
24. Narayanan VA (2005) Ethanol alternative, *The Indian Express*, 11 September 2005
25. Pulakkat H (2010) Next gen biofuels: clean power, *Businessworld*, 31 May 2010, p 79
26. Rajagopal D, Sexton SE, Roland-Holst D, Zilberman D (2007) Challenge of biofuel: filling the tank without emptying the stomach? Environ Res Lett 2:1–9 IOP Publishing Limited, UK
27. REA (2006) South African company to build $1b ethanol plants, *Renewable Energy Access*. http://www.renewableenergyaccess.com/rea/news/story?id=44332. Accessed 13 March 2006
28. van Eijck J, Romijn H (2008) Prospects for Jatropha biofuels in Tanzania: An Analysis with Strategic Niche Management. Energy Policy 36:311–325.
29. Viswanath R (2006) Alternative energy: sweet smell of ethanol, *Businessworld*, 12 June 2006, p 42
30. Widenoja R, Halweil B (2008) Analysis: banning "bad" biofuels, becoming better consumers. Eye on earth. Worldwatch Institute, Washington
31. Worldwatch (2008) Grain harvest sets record, but supplies still tight. Vital signs 2007–2008. Worldwatch Institute, Washington
32. Yusoff S (2006) Renewable energy from palm oil—innovation on effective utilization of waste. J Clean Prod 14:87–93

Chapter 5
Mapping Development Policy onto the Life-Cycle

> *You have to do your own growing no matter how tall your grandfather was.*
> Abraham Lincoln
> 16th President of the United States of America

Financial support and other incentive structures need to necessarily mirror the maturity of the industry segment under consideration. As a market matures, service provision and financing are likely to represent significant revenue opportunities while dwindling margins on module/equipment manufacture would expedite formation of vertically integrated energy service delivery chains. Following the discussion on the power of scarcity, and the role played by substitutes, we apply the product life cycle framework to analyze the impact of global trends on the photovoltaic industry and for the solar thermal / water heating sector. End-users would be required to pay for the power/service alone while energy service providers would own and operate the generation equipment, much as utilities own generation assets in centrally generated, grid-supplied power systems. The first part of this chapter discusses global trends in solar PV module production and attempts to chart a course for the Indian industry. The latter sections deal with appropriate financial support to help the transition from one stage in the life-cycle to the next, employing the solar thermal industry as an illustrative case.

The Renewable Energy sector[1] in general and Solar Photovoltaic (SPV) technology in particular are often labeled the "sunrise" industry and are said to be in their 'infancy', but, over time, economists have charged it with being in this state for too long, [14]. Solar PV's promised drastic cost reductions have not materialized [6, 37] and potential market sizes have been grossly overestimated, according

---

[1] This chapter is largely based on Srinivasan, Sunderasan, "The Indian Solar Photovoltaic Industry: A Life Cycle Analysis," Renewable and Sustainable Energy Reviews, Vol. 11, No. 1, Elsevier Ltd., January 2007.

S. Sunderasan, *Rational Exuberance for Renewable Energy*,
Green Energy and Technology, DOI: 10.1007/978-0-85729-212-4_5,
© Springer-Verlag London Limited 2011

to Bradley Jr. [7]: paradoxically, with the German 'roof-top' program market volumes grew rapidly while prices began to rise owing to supply side constraints. Conversely, with the collapse of the Spanish program in 2008–2009, manufacturers were left with high levels of finished goods inventory. One general observation is that the industry has grown on the back of grants and subsidies and has been accorded special treatment the world over, especially given the heightened fears relating to climate change and global warming, or merely to create additional jobs.

Yet, it is time to subject the solar PV industry to rigorous economic analyses alongside other 'normal' industries. Apart from the social and ecological aspects of growing the solar PV industry, a strategic approach needs to be adopted to identify the financial and economic barriers [43], causing the gap between the actual and the potential deployment as a starting point to designing innovative energy policies and to creating an enabling environment [57]. Polatidis and Haralambopoulos [40] have found that a long–term normative perspective for renewable energy decisions covering political, legislative, administrative, economic and marketing issues is an important prerequisite for their successful deployment. This chapter is an attempt to apply the industry life-cycle theory to the (Indian) Solar PV industry, focusing predominantly on the solar module manufacturing operations, and to chart a course for its evolution beyond the protectionist and subsidy era.

From the perspective of Japanese, European and American governments, beyond export markets and technological leadership, the prospect of reduced dependence on imported oil, lower air pollution, obviating nuclear safety risks, and climate disruption provide justification for the deployment of micro-power technologies such as SPV, [39]. In India as in other developing countries, lowering the need for the extension of the utility power grid and limiting transmission and distribution losses and theft of power for the electricity utilities are additional drivers for government policy.

5.1 The Product Life Cycle

Most industries evolve and go through a life cycle, typically mimicking the biological life cycle. Klepper and Graddy [22] have reported the regularity of the product life cycle (PLC) of 46 products. The industry comes into existence and begins to grow in stage I, often as a result of an innovation. During stage II, referred to as the shakeout phase, companies exit at an even faster rate than their entry and Stage III refers to maturity of the industry, followed by its eventual decline and replacement. A general consensus exists among authors that once a product is launched, imitators appear rapidly and in great numbers, expanding output dramatically, resulting in falling prices, [54]. The market then selects product designs and R&D tends to shift from an emphasis on improvements in product design to improvements in the production process itself. Firms with superior production and distribution and relatively greater improvements than the competition survive the shakeout. Once the industry matures, the industry concentration does not change dramatically, while R&D is concentrated on cost reducing innovations and minor improvements in product design. Some products suffer technological obsolescence and, in due course, die out, [34].

The PLC approach has been applied to various settings, from the hitech commercial mainframe computer industry [13] to unsophisticated cowpea based agricultural products, [36]. Werker [55] observes that "a crucial characteristic of the product life cycle approach is that markets change from being more favorable for entrants to being favorable to established firms". Karlsson and Nystrom [21] have found evidence of differences in knowledge intensity for firms in different stages of the product life cycle.

Jovanovic and MacDonald [18] have attributed shakeouts to major technological innovations more or less exogenous to the industry. In the present context, we could consider large scale success with thin-film technology (favorable) or commercially successful alternatives, viz., fuel cells (unfavorable) leading to a shakeout. More relevant and immediate, however would be cost advantages and economies of scale [23] in enhancing efficiencies of the production process. It has been observed that the solar PV industry has settled on the end product (predominantly crystalline silicon) and cost differences dominate product differences. According to Jackson and Oliver [16], it is possible to "envisage a virtuous circle of market growth, expanded production and further economies of scale" in the near term.

5.2 Background and Industry Global Trends

The world-wide solar PV industry had seen dramatic growth rates of 27% over the five years to 2004, even reaching 32% yoy in 2003 when 742 MWp[2] of cells were sold, taking the cumulative world production to 3145 MW, [17]. The global solar industry had grown to $7-billion per annum, largely on the back of government incentive programs, together with lower prices secured through volume purchases. The industry had experienced an annual growth rate in excess of 18% in over the previous decade, [38]. Yet this pales in comparison to the explosion witnessed in the latter half of the same decade. It is estimated that over 7300 MW of solar PV was installed in 2009 alone (almost ten times as much as 2003 volumes), on the back of 6000 MW of installation in 2008. By early 2010, the global installed solar PV capacity had topped 21,000 MW. At 3,800 MW, Germany alone was responsible for half the 2009 global solar PV installation volumes, while Spain dropped from a very respectable 2700 MW in 2008 to a lowly 70 MW in 2009, in tandem with sharply reduced feed-in-tariffs. Italy (580 MW), Japan (480 MW), the United States (470 MW) and the Czech Republic (410 MW) followed Germany on the leader board, [42].

---

[2] MWp = mega Watt peak and kWp = kilo Watt peak.

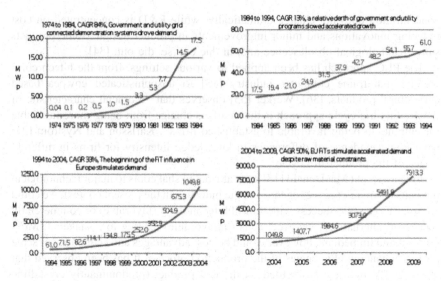

Fig. 5.1 Growth of the solar photovoltaic market (1974–2009)

Figure 5.1 from Mints [29] offers a history of PV industry growth from 1974 through 2009 broken into 10-year periods, highlighting the explosive growth in the five years to 2009.

Since the early days, the industry has depended on fiscal incentives and demonstration projects for its growth. Ironically, the stunning growth witnessed in the first decade of the century is also on the back of substantial preferential-tariff schemes in Japan, Germany, Spain, Italy, the United States and a few other markets. Such dependence on government schemes has also meant that the solar PV industry has had to grapple with the uncertainty associated with continuance of promotional programs. The growth is therefore not monotonic but a series of ups and downs, as illustrated in Table 5.1 [29].

This has resulted in the progressive reduction in module prices as displayed in Table 5.2 [30], owing to scale economies, more efficient production techniques and to incremental cell/module efficiencies.

Traditionally, PV prices have been projected to reduce by 5% per year [26] to the year 2010, or by 20% with experience gained with every doubling in installed capacity, [12, 17]. In reality, the prices for PV modules prevailing in the international markets are determined by the dynamics of demand and supply, launch of subsidy programs in different countries and the production volumes and efficiencies of relatively large manufacturers. The general observation is that producers have been progressively ramping up production capacities to meet projected prices and no single producer might be in a position to exercise substantial market power, even in what, at first sight, appears to be an oligopolistic setting. This could be attributed to the nature of the industry and its current level of maturity, given that demand is largely driven by subsidy programs and other government and donor initiatives and procurement is generally managed or supervised by industry experts and consultants.

Years	MWp	Change (%)
1974	0.04	
1975	0.1	150
1976	0.2	141
1977	0.5	87
1978	1	112
1979	1.5	53
1980	33	128
1981	5.3	59
1982	7.7	45
1983	14.5	88
1984	17.5	21
1985	19.4	11
1986	21	8
1987	24.9	19
1988	31.5	27
1989	37.9	20
1990	42.7	13
1991	48.2	13
1992	54.1	12
1993	55.7	3
1994	61	10
1995	71.5	17
1996	82.6	16
1997	114.1	38
1998	134.8	18
1999	175.5	30
2000	252	44
2001	352.9	40
2002	504.9	43
2003	675.3	34
2004	1049.8	55
2005	1407.7	34
2006	1984.6	41
2007	3073	55
2008	5491.8	79
2009	7913.3	44

Table 5.1 Annual growth of the solar PV industry (1974–2009)

While the PV industry was regarded as being in the early part of the learning process during the decade of the 1990s [37], clearly, in recent years we observe the signs of a maturing industry. More obvious is the evidence of the shake-out and consolidation phase in the industry—the increase in market concentration and the squeezing out of relatively smaller players or non-specialized players. In early 2002, Shell Renewables bought out a joint venture with Siemens and E.ON to become one of the larger PV companies in the world. This has been hailed as "good news for customers, staff and partners, and for the whole solar industry," as the "integrated

Table 5.2 Solar PV module average selling prices (ASP) in 2009 $/Wp[a]

Year	Smalt-quantity buyers		Mid-range buyers		Large-quantity buyers	
	ASP $/Wp	Change (%)	ASP $/Wp	Change (%)	ASP $/Wp	Change (%)
1994	$5.39	−1	$5.29	−1	$5.19	0
1995	$5.11	−5	$5.01	−5	$4.90	−6
1996	$4.44	−13	$4.30	−14	$4.15	−15
1997	$4.40	−1	$4.29	0	$4.18	1
1998	$3.79	−14	$3.74	−13	$3.55	−15
1999	$3.80	0	$3.77	1	$3.30	−7
2000	$3.82	1	$3.79	1	$2.75	−17
2001	$3.57	−7	$3.50	−8	$2.65	−4
2002	$3.31	−7	$3.25	−7	$2.75	4
2003	$3.14	−5	$3.10	−5	$2.65	−4
2004	$3.65	16	$3.35	8	$2.90	9
2005	$3.85	5	$3.65	9	$3.03	4
2006	$5.00	30	$3.90	7	$3.39	12
2007	$5.10	2	$3.98	2	$3.50	3
2008	$5.02	−2	$3.65	−8	$325	−7
2009	$3.68	−27	$2.82	−23	$2.18	−33
CAGR 1994–2004	–	−4	–	−4	–	−6
CAGR 2004–2009	–	0	–	−3	–	−6
2009 ASP	€ 2.59	–	€1.98	–	€1.53	–

[a]Module average selling prices (ASP) for large-quantity buyers (typically 50 MWp to > 100 MWp a year) fell 33% from $3.25/Wp in 2008 to $2.18 Wp in 2009
Module ASPs for mid-level buyers (typically 10 MWp to > 25 MWp per year) fell 23% from $3.65/Wp in 2008 to $2.82 Wp in 2009
Module ASPs for small-quantity buyers fell 27% from $5.02/Wp in 2008 to $3.68/Wp in 2009

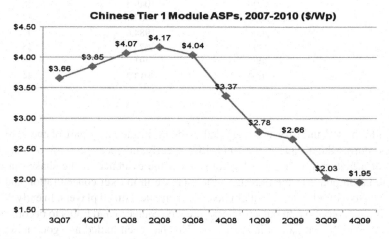

Fig. 5.2 Average selling prices for Chinese solar PV modules

Table 5.3 Growth of the solar photovoltaic shipments

Region	2004	%	2005	%	2006	%	2007	%	2008	%	2009	CAGR
US	140.6	−5	133.6	2	136.6	74	237.3	63	387.9	5	409	24
Japan	547	31	714	24	882.6	2	901.9	36	1228	1	1240.4	18
Europe	272.9	49	406.9	50	611.3	62	992.4	71	1700	−13	1487.2	40
ROW[a]	47.2	45	68.6	57	107.8	49	160.5	162	421.2	164	1113.8	88
China/Taiwan	42	101	84.6	191	246.3	217	780.9	125	1754.7	106	3610.9	144
Total shipments	1050	34	1407.7	41	1984.6	55	3073	79	5491.8	43	7861.3	50

Shipments (in MWp) from China/Taiwan, Europe, Japan and ROW 2004–2009
[a] ROW region includes India, Malaysia and The Philippines, among other countries. Most other countries have facility in pilot or pre-pilot stage

company will have the people, the reach and the resources to build a sustainable, commercially successful solar PV business around the world" [47]. Progressively, diversified energy players have yielded market space to specialist firms engaging exclusively in the manufacture of solar PV equipment, viz., Q-cells of Germany, or providing installation and post-installation maintenance. Shell Renewables, for instance has restructured operations and has moved out of the PV industry. The genuine black swan[3] in the solar PV industry has been the rise of Chinese manufacturing, rapid scale-up and decline in production costs as in Fig. 5.2, [56], with shipments from China and Taiwan accounting for nearly half of the global total as shown in Table 5.3 [31]. The growth in volumes and the lower costs are attributed to the Chinese equipment makers' control of the entire supply chain, "from raw material through to wafer, ingot and cell manufacturing, module assembly and in some cases installation".

In recent times, the solar module has been 'commoditized', with producers competing almost exclusively on cost while technology, branding and country-of-origin are no longer significant distinguishing factors. German solar cell manufacturer, Q-cells for instance, has taken to manufacturing of crystalline solar modules and to installing mid-sized solar systems ranging from 5 kW to 5 MW for industrial clients. Simultaneously, the firm is slated to move jobs to its production facility in Malaysia, signaling perhaps the saturation of the German market, and the creation of opportunities in the emerging Asian markets, [50].

5.3 Projections for the Indian Solar PV Industry

Given this background of increasing concentration and declining prices, we are now in a position to investigate the impact of these global trends on the Indian solar PV industry. We could study the consequences of the shake-out and industry maturity on Indian units and identify their sources of comparative advantage.

---

[3] Random events that are nearly impossible to predict; leaving huge impacts on our lives, once they occur.

Proposition 1 *Consolidation and shake-out in the Indian PV module industry is imminent and may be considered almost inevitable. Entrenched players who have amortized their first costs are likely to survive in the face of declining product costs while new entry by small/medium scale players would be discouraged.*

In the early part of the decade, the Indian solar photovoltaic module manufacturing industry consisted of over 20 companies with aggregate production capacity of just over 70 MWp, [44, 52]. This included the capacity of one large player, with almost 55% of the industry capacity. The remaining part of the industry capacity was divided among several government, joint sector corporations and private companies. A large number of the fringe module manufacturers sold complete lanterns and PV systems and services rather than modules by themselves.

Global players have invested heavily in automated facilities and large capacities, while, in contrast, capacity addition by most fringe Indian producers has been marginal and capacity utilization, modest. Among the larger players, TATA BP Solar, a well entrenched player and long-time industry leader is home to 125 MW of module capacity and about 84 MW of cell capacity. Even as the company has been exporting over 70% of its produce, it hopes to supply substantial quantities of modules to the Indian Government's of the proposed National Solar Mission,[4] [51]. One of the other larger players, albeit a late entrant, Moser Baer, an Indian optical disk maker had sunk close to USD 180 m, had mobilized an additional USD 90 m as of mid-2009, in its quest to invest close to USD 1.5 billion in the solar PV sector. The company has set up a 80 MW crystalline silicon production facility and a 40 MW thin-film facility in northern-India and was scheduled to grow capacity to 600 MW. In contrast to TATA BP, given adverse market and credit conditions, and in the face of piling inventories and declining finished-product costs, Moser Baer had to scale-down proposed investment for the year 2009 from about USD 185 m to just over USD 20 m. Additionally, polysilicon feedstock that goes into the production of solar cells has been in short supply owing to imbalances in global supply chains, [32]. A classic illustration of a maturing industry supporting an established incumbent, at the expense of entrants.

With finished product prices declining rapidly, and with little or no difference in technology within the industry, like most other matured industries worldwide, we can expect that with increasing scale, a firm's marginal cost is lower than its average cost of production. The Indian industry is bound to experience progressively rising real and nominal labor costs, and, in the face of declining product prices, margins could be sustained only if the Indian producers scale up rapidly to derive the benefits of scale and increase capacity utilization approaching global standards, while striking a balance between automation and application of labor. Few entrepreneurs and business groups would be in a position to mobilize additional investments, especially in the face of tight global financial markets. Suboptimal plants can exist for brief periods, by doing things differently [5], but are eventually likely to be overwhelmed by developments. By 2007, analysts had

[4] http://mnre.gov.in/pdf/mission-document-JNNSM.pdf.

projected aggregate Indian industry production capacity to top 120 MWp[5] but substantially more concentrated, with an average production capacity far greater than the mean of about 3 MWp observed in the year 2000. Some of the players have since terminated cell manufacturing and have restricted themselves to larger volume module operations using imported cells or cells from other Indian manufacturers. With declining tariff and other barriers, the smaller players would have to migrate to service delivery or exit the industry completely. This is in also line with Hoffman's findings [15] that, renewable energy/energy efficiency businesses, upon deregulation, would ultimately consolidate into fewer and larger firms.

Presently, production at some plants occurs in spurts, following launch of subsidy programs in India or overseas, while average capacity utilization over time remains moderate, and inventory build-up (and shortages) are cyclical, corresponding with the ebbing and rising of subsidy programs and preferential tariffs. Organized marketing efforts from larger producers and powerful brands would be required to enhance and even-out production. The larger, well-entrenched players are likely, also, to be better placed to enhance capacity utilization, by promoting their wares more aggressively and by planning for a longer-term, attracting professional management talent, requisite investments and timely working capital.

Proposition 2 *Vertical integration with service providers necessary to compete in the domestic consumer markets. Domestic as well as overseas equipment manufacturers would have to partner with local service providers to tap into retail demand as well as the demand from the proposed National Solar Mission.*

Solar home systems (SHS) have by far represented a large proportion of the rural electrification market that has used PV systems.[6] Certain parts of the country have witnessed steady volume growth on the back of government sponsored capital subsidy programs, and others have made extensive use of third party credit from commercial banks, rural cooperatives etc. Invariably these markets are served by system integrators and service providers such as SELCO India, Shell Solar India (now acquired by an Indigenous energy service provider), TATA BP dealerships, Environ Energy Tech Systems (Calcutta) and NGOs and social organizations such as the Ramakrishna Mission (RKM) and the Social Work Research Center (SWRC). These integrators are typically, local or regional players and benefit from an in-depth knowledge of the socio-economic-political realities in their geographies and enjoy a good rapport with their customer base and with financial intermediaries.

In India, Renewable Energy Service companies and PV systems are not expected to be substitutes for the electricity grid in the near-term. These mini-utilities help reduce peak loads, reduce transmission and distribution losses and help optimize grid extension and usage and we could expect them to play a more

---

[5] Assuming an annual increase in capacity of about 20% in line with global capacity expansion.
[6] For an extensive discussion on the segments of the Indian PV market, please refer to Srinivasan [49].

formal role in the near future, [27]. PV brings the service closer to the ultimate consumer, as Seetha [45] puts it, water for the farmer, light for the family and better indoor air quality for the entire household. Improved competition among these solar utilities ultimately benefits the customer. The business risk for the service companies, themselves, is reduced by the smaller capital outlays required and the shorter gestation periods.

Each PV service provider typically sources entire systems or components of systems from established manufacturers. These service providers, essentially, represent the domestic distribution arms of the module and component manufacturers, and are nimble enough to provide the grass-roots level customer interface and periodic service. As the market has grown and matured, most module (and other component) manufacturers realize that it is neither their core competence nor are they appropriately placed to provide these "energy services" to the end customers, themselves. On the other hand, some of these dedicated service providers or energy service companies have achieved sufficient critical mass of installations, reputation for service and the momentum required to expand rapidly.

The urban power pack market would be opened to the PV industry as soon as banking with the utility grid and net-metering are permitted, as the operational economics are already rendered comparable to the ubiquitous Diesel generator, [24]. The revenues accruing from the sale of the ecological benefits from these systems would expedite convergence with the grid or other possible alternatives such as inverter based uninterruptible power supply systems (UPS). As the Worldwatch Institute puts it, "New business models would have to evolve around the micro-power technologies" [39], and module manufacturers would attempt to lock-in the service providers. In the next phase of consolidation and reorganization, and with further maturing of the industry, *energy service provision and financing would represent the predominant revenue opportunities* and manufacturing of (commoditized) modules and components would drift to the background, with minimal profit margins (zero 'economic profit').

Subject to the legal stipulations, and with a view to tapping the opportunities emerging from the proposed National Solar Mission, the module manufacturers (domestic or overseas) would be encouraged to take equity positions in the existing/domestic energy service companies. Each manufacturer is likely, therefore, to be a minority equity holder[7] in a few energy service companies operating in different geographies. This equity infusion also strengthens the balance sheet of the service providers and provides them with much-needed liquidity, while assuring them of uninterrupted module supplies. The consequent structure of the reorganized energy service delivery value chain is shown in Fig. 5.3, and analysts expect to witness the development of this structure commencing forthwith and expanding continually as the National Solar Mission begins to take root.

---

[7] Sufficient to ensure module offtake without management responsibility.

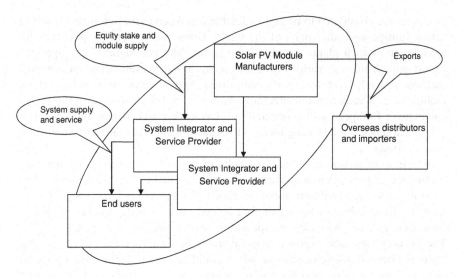

Fig. 5.3 Evolving structure for solar photovoltaic energy service provision

Proposition 3 *Comparative advantage for Indian manufacturers would reside in the export of small modules, (< 70 Wp), and the Indian market would benefit with the import of larger modules* (>100 Wp).

Over the first half of this decade, an Indian business group attempted to import thin-film modules, which are, in theory, cheaper than crystalline modules, but found that the transportation costs and customs tariffs thereon rendered them price-uncompetitive within India. During the same period, a European firm attempting to export solar lanterns in a knocked down condition to India, for subsequent assembly and retail sales, faced similar cost barriers. The situation is set to change rapidly with simultaneous decline in international product costs and tariffs imposed thereon. Further, with the prices of crystalline silicon cells declining rapidly, thin-film modules find it more difficult to compete, given their relatively lower efficiencies and larger sized installations.

When benchmark international prices fall below the production-costs of the Indian manufacturers, progressively, the smaller players in the industry cannot exist without tariff protection. We would soon encounter the situation where the landed cost of imported modules would be lower than the average cost of production at a small facility in India, owing to the higher production efficiencies and scale economies accruing to large automated facilities abroad. At this stage we are likely to see imported modules entering the Indian market. The Indian units could partner with the larger overseas players to add 'local content' and/or to provide installation and maintenance service.

Exports are an integral part of the industry's growth and India is increasingly seen as a low cost manufacturing base for high-quality small modules (< 70 Wp). Since most module manufacturing operations tend to be labor intensive, Indian producers are considered to be highly flexible compared to the automated plants in Japan, Europe and other parts of the world. Consequently, energy paybacks for modules from such plants are also far superior. The government has supported export growth through various measures including duty exemptions on imported material (intended for re-export), subsidized interest rates on working capital facilities etc. The Indian manufacturers have successfully exported to the highly competitive European grid-connected/roof top market and African rural electrification markets and are looking forward to expanding into Australia and South-East Asian nations.

By virtue of producing "small" modules of varying sizes through flexible production techniques, often supported by international quality certification, the Indian exporters have managed to negotiate marginal premiums over international prices. Export related subsidies have therefore helped create a niche market for Indian manufacturers, a niche which is expected to grow as these export markets evolve. The industry views the export market favorably because it increases capacity utilization, brings in foreign exchange which could be used to pay for silicon wafer/ cells and other imports, and because of its importance as a hedge against sluggish demand growth in the domestic market, and other such disruptions.

Proposition 4 *Imported solar modules would increasingly find their way into the rapidly growing commercial markets.*

Corporate houses in India have rolled out networks of internet kiosks in rural and semi-urban areas to enhance information flow with these areas, help enhance farm productivity, improve farm-gate price realization and cut transaction costs. Agriculturists are now in a position to access latest and relevant information on weather, better farming practices as well as market prices, through specially designed internet portals—all in their respective vernacular languages. These kiosks also facilitate supply of farm inputs as well as purchase of the farm produce, [48]. Kiosk networks are also operated by provincial governments to provide e-governance services. Since most of these kiosks are located in areas with indifferent power supply, PV systems are the obvious source of power to operate these kiosks. Private telecom companies rolling out WLL[8] and other rural telecom services and have been ramping up their consumption of PV modules for powering network interface units, base, booster and repeater stations. Each of these market segments is on the verge of an explosion, given the emphasis placed on the use of ICT[9] to bridge the urban–rural divide.

Similarly, the market for cathodic protection, signaling etc. is expected to grow exponentially as networks of oil and gas pipelines, highways etc. are being developed by private players. Voice and data communication, signaling systems, cathodic protection and other miscellaneous applications thus, represent an obvious and large nation-wide market for PV systems. Most of these PV systems are

---

[8] Wireless in Local Loop.

[9] Information and Communication Technology.

designed for the specific application in question and are essentially packaged with the appliances being powered. Service arrangements for these systems tend to be clubbed with the maintenance support for the rest of the hardware—the computers, printers, telecom equipment and other end use appliances.

The business model for commercial applications of PV modules (including power plants installed under the ambitious National Solar Mission) would therefore represent design and delivery while maintenance could, typically, be carried out by third parties, viz., local partners. This market could be viewed as being competitive insomuch that price, quality, reliability, brand equity and risk of delivery default would play a role in the procurement decision. Intra-industry differences in prices could be substantial, depending on the nature of the service arrangement sought by the client and the geo-spread of the installations.

Commercial markets are expected to see explosive growth, especially if things go as planned under the National Solar Mission. Post-installation service could be provided by third parties, with or without previous experience in the industry. Systems are likely to be sold as an integrated kit along with the application in question—internet kiosks, signaling equipment etc. The imported modules, especially > 70–100 Wp, are likely to become much more affordable with production costs at large overseas facilities falling relative to Indian costs, simultaneous with falling tariff barriers. This market segment would be willing to pay a premium for superior technical design and aesthetics and is therefore likely to be the first to witness the entry of imported modules, essentially packaged with the end-use application, viz., automated teller machines (ATM).

Proposition 5 *Government (supported) programs would continue to absorb sizable output volumes but would need to initiate confidence building measures.*

Power plants especially those proposed under the National Solar Mission, could vary in size from a few kWp to over several MWp and thus represent a very attractive opportunity to place sizable volumes of modules on the market in one go. And yet, these plants are often installed in remote areas, arid lands, islands and mountainous regions, and consequently, it is quite challenging for the module manufacturers to provide post-installation service. The most common way around the qualification criteria, the service-provision mandate and the working capital requirements, is to partner (viz., by forming special purpose joint venture companies) with local integrators/service providers who actually install and service the plant while the module manufacturers lend their names, supply solar PV modules and get paid upfront. Service provision in case of pumping systems is slightly more intensive and is of the highest intensity for the street lighting systems as the system sizes decrease and spatial dispersion is vastly increased. The local players would also serve as agents in negotiating the bureaucratic process involved at each stage of installation and operation.

Global production capacities have risen rapidly relative to increase in production capacity by Indian manufacturers. Despite the protectionist tendencies and the lobbying of domestic producers, the demand slated to be generated by the proposed Solar Mission is unlikely to be met exclusively by domestic producers.

Large global players, with relatively lower costs of production (for the large modules required for such power plants) would definitely be required to participate. However, the indifferent financial health of the electricity utilities and the long-drawn legal procedures could still dissuade serious investors.

In addition to announcing aggressive tariffs and ambitious targets, the government would be required to create confidence among industry players. For instance the Mission provides for a periodic review of the preferential tariffs offered for solar power generation projects. Such uncertainty would be unacceptable to most equity investors. Further, mechanisms need to be put in place to ensure that payments for power generated are made within acceptable time frames. A suitable appeals process (not requiring to go through the clogged judicial-system) would also need to be instituted to attract professional investors.

5.4 Summing Up

The three segments of the Indian PV module market, namely the government, commercial and consumer segments are likely to grow in different proportions. Government procurement would remain steady and be dictated more by political convenience, rather than with industry evolution in mind. In order to serve the consumer better, a level playing field should be provided for domestic and imported equipment to compete on the basis of cost and quality. The qualification criteria need modification to permit the participation of professional system integrators and service providers directly (with assured supply of modules, electronics and other technology from reputed domestic/global players) in the tender process, especially for power plants under the National Solar Mission.

Smaller module makers with existing service operations would tend to swap their production activities for enhanced service opportunities. The industry would therefore comprise a few entrenched module manufacturers, generally utilizing imported cells, each with equity positions in several mini-utilities/service providers. The market is likely to grow on the back of commercial finance and systematically paid preferential tariffs, and on the promise and evidence of disciplined post-installation service.

Consequently, financing and market development programs in India and other similar settings should be targeted at facilitating this consolidation in manufacturing operations and enhancing quality standards for the service operations and most importantly for attracting more banks and financial institutions into working capital and end-user financing for solar PV. Government policy should be directed towards making the industry competitive [58] in the international arena, and eliminating the perpetual need for financial assistance or concessions of any sort. Subjecting the photovoltaic industry to rigorous economic analyses alongside mainstream industry marks a turning point and provides policy makers with an opportunity to look at it differently, and restrict subsidies and preferred treatment to genuinely less-developed areas and under-privileged populations.

5.5 Policy Support Over the Lifecycle

This section[10] tracks the supportive measures necessary to incubate, sustain and grow the renewable energy industry at various stages of maturity. Appropriate financing at various stages of maturity is seen as a key driver for growth and consolidation. Policy measures to mainstream the industry as it attains critical mass are prescribed, employing the solar thermal water heating sector for illustration.

Solar water heating systems (SWH) replace electric/LPG geysers, firewood or other fuels, appliances and techniques used to heat water, and hence contribute to energy conservation. In addition, large scale deployment of the SWH helps reduce peak load demand for the electric utility. Since they are located on rooftops, SWH conserve indoor space and enhance safety in the wet areas by preventing electric shocks, short circuits etc. SWH also reduce the consumption of fuels such as firewood, coal, furnace oil etc., and contribute to mitigation of CO_2 emission and reduction in degradation of the environment. SWH have healthy monetary paybacks and with an LPG or electrical backup for non-sunny days, can provide heated water round the year. Despite the highly visible and readily acknowledged advantages of the product/technology, actual global deployment pales in comparison to potential.

The most common active solar technologies for heating water are the flat plate collectors and the evacuated glass tube based systems, each working on the principle of the thermo siphon. Chapman [9] has applied the discounted cash flow analysis to evaluate the cost-effectiveness of solar thermal systems in domestic dwellings. Clearly, the product has evolved since then. It is estimated that a 100 liter per day (lpd) capacity SWH saves approximately 1500 kWh of electricity annually, [33] and could prevent emission of 1.5 tonnes of CO_2 per year. The monetary payback could range from 2 years if electricity is replaced through 7 years if coal is displaced.

5.5.1 Relevant Experiences

The Greek solar thermal market is one of the most developed markets worldwide. Karagiorgas et al. [20] have evaluated solar thermal installations in Greece in economic terms in comparison with energy equivalent systems such as LPG, natural gas, diesel etc. Kaldellis et al. [19] have advanced this line of thought by undertaking a cost-benefit analysis in an attempt to find a rational explanation for variation in demand patterns across time. The REACT [1] case study on the Greek domestic

---

[10] This section is largely based on Srinivasan, Sunderasan, "Transforming Solar Thermal: Policy Support for the Evolving Solar Water Heating Industry", ReFocus, Elsevier Ltd., March/April 2006.

SWH market and Argiriou and Mirasgedis [4] attribute the success in large-scale deployment to technology reliability and installation and service quality.

Rapid expansion of the Chinese economy has assisted the development of the RE industry over the past two decades. Xiao et al. [59] have reported successful examples of solar thermal utilization in China and have identified barriers to large-scale deployment. Muneer et al. [35] have evaluated the prospects for the SWH in the context of the Pakistani textile industry and have arrived at monetary paybacks of slightly over six years with energy paybacks of just about half a year.

Ardente et al. [2] have assessed the energy balance between the energy consumed and energy saved during the life-time of the solar thermal collector in Italy. Over and above the monetary saving accruing from displaced grid electricity, the SWH contributes to a healthier environment. Ardente et al. [3] have applied the ISO 14040 international standard to trace the solar thermal system's eco-profile to synthesize the environmental impact related to the entire life cycle, commencing with manufacture, through various phases of installation and maintenance and final disposal. The energy and environmental payback from the solar thermal water heater is obviously very attractive, *relative* to the nearest potential substitute, viz, the electricity driven storage water heater.

Besides Greece, India, China, Egypt and Turkey are among the major markets for solar water heating systems, [53]. Shivakumar [46] encourages the use of building-integrated SWH, built into the roofs of residential units. Chandrasekar and Kandpal [8] have estimated the potential number of Indian households with the capacity to invest in SWH, given the capital cost, cost of loans and the demand for heated water. In addition to domestic application, large scale heating systems are installed at boarding schools, dairies, textile units, chemical and bulk drug units, breweries, large hotels, hospitals and other institutions. Their deployment in these situations replaces the use of conventional appliances such as storage water heaters, wood-fired water boilers etc., and saves costs in these industries and institutions.

As the markets for SWH mature and volumes and installation density attain significant proportions, emission reduction revenue from the sale of Certified Emission Reductions (CERs) could make meaningful contribution to the projects involving SWH technology, [28]. These accruals could be deployed on an industry-wide scale to develop other markets, create awareness on the cost-benefit dynamic of SWH, help create capacity to improve installation and service quality, and to overcome other barriers that constrain widespread dissemination.

5.6 Stages of the Solar Water Heater Life-Cycle

While the industry has taken off in some countries, the global potential for deployment is under-harnessed, [11]. A number of explanations have been advanced including: the lack of purchasing power, the non-availability of running water, climatic conditions incompatible with the utilization of hot water and the fact that domestic application of hot water is non-income generating. Winkler [57] examines policy options for promoting renewable electricity in South Africa and the analysis could be extended to include other non-conventional energy technologies. He opines that an enabling environment is created by allowing renewable energy technologies to compete on a level playing field with alternative options.

With a technology that is technically established and commercially viable, policy is to be directed at creating an enabling environment which has a greater impact than merely offering direct financial support. A primary economic instrument which policy makers would be tempted to employ would be to tax conventional devices such as storage water heaters, and offer tax breaks to promote the "green" alternative. In the Indian context, where subsidy on power is a politically sensitive issue, attempts have been made to offer balancing subsidies on alternatives, [10]. Market driven instruments such as "green" certificates to give credit to the power saving from SWH have also been tested, [41]. Ministries and implementing agencies have also attempted to make the installation of SWH systems mandatory in certain categories of buildings. Some utilities have offered discounts on electricity tariffs for residences installing SWH, an apparently counter-intuitive measure as the rebate needs to be offered on the consumption avoided. Worse still, the effect of such rebates would depend on the relative proportion of water heating loads to other loads. Predictably, these policy measures applied individually or in combination, have yielded mixed results, at best. *The appropriateness or otherwise of a certain policy initiative depends on the relative maturity or stage of the life-cycle of the industry in discussion.*

The SWH industry, like most others, goes through several distinct stages in its evolution [25]. The product or technology is incubated and tested at universities, laboratories and test facilities and set out for demonstration at high-visibility sites. The particular technology selected depends on the availability of raw material, viz., copper, glass etc. in this case, the solar insolation and the financial standing of the target market segments. This incubation stage is followed by the set-up of test protocols and standards to be followed in manufacture and installation. Simultaneously, indigenous capacity in these areas is built-up. As the technology approaches adolescence and its applicability in the given context is established, the industry is invited to take the product closer to the end-markets. During each of these stages, a tremendous amount of hand-holding is required by the policy making/apex body with the help of national and international experts. As the industry gets to grips, and the product lends itself to mass production, several newcomers enter the industry and generally replicate the incumbent's product and processes. At this stage, the product support from the apex body needs to give way to a policy framework which encourages competition and ensures quality of installation and promptness of service. Pollution standards, especially at the manufacturing facility are to be imposed to ensure disciplined treatment of wastes generated.

Concomitantly, the nature and quantum of financing undergoes changes during this evolutionary process. The demonstration of equipment and transfer of expertise are almost always grant funded. Industry initiation is accomplished by subsidized funding while mature technology is sustained by fixed and working capital from mainstream sources. Further expansion is determined by this

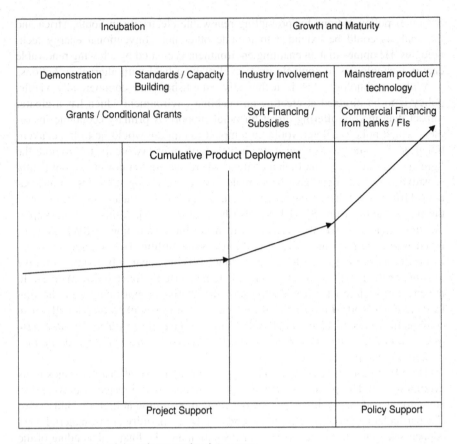

Fig. 5.4 Life cycle of the solar water heating system

availability of such capital which also leads to industry consolidation and scale economies. Also as the industry matures, the product is standardized and product improvements are dominated by process efficiency improvements. This stage-wise analysis is summarized in Fig. 5.4.

5.7 Project vs. Policy Support

The early stages of the SWH lifecycle require hands-on *project* support from the implementing agency.

- Incubation is achieved by facilitating technology and expertise transfer. Product demonstration and standardization are followed by awareness creation and communication of the beneficial aspects of the technology.
- As experience from the Greek market suggests, creating a favorable first impression for product reliability is vital and this is achieved through stringent

quality checks, adequate design and proper installation supported by disciplined service. For instance, insufficient insulation of the hot water tank alone could lead to rapid temperature drops in the heated water, and consequently to lack of confidence in the efficacy of the technology itself.

- Fiscal incentives such as tax holidays and other concessions help involve industry. Simultaneously applied capital subsidies on equipment help create a market and to absorb the overheads associated with small production volumes. Subsidized funding at this stage also supports the set-up of service delivery chains and for large-scale capacity building.

All of the above require budgetary support or donor funding or a combination thereof. As the product is standardized, it is vital to move from a supply-push to a demand-drive. This is facilitated by the provision of consumer finance to make the product more affordable. *Appropriate financial facilitation is vital to ensure prompt and smooth transition from one stage to the other.* Making subsidized funding available to a mature market is as superfluous and counter-productive as providing consumer finance at the product demonstration stage is futile.

A simultaneous switch needs to be made from *project to policy* support. The ministries and other enabling agencies need to distance themselves and permit the "invisible hand" of the market to take control. With adequate product experience and resulting brand recognition, the market is empowered to reject poor quality. Process improvements deliver increased efficiencies up to a certain level, beyond which, scale economies are achieved by acquisitions and consolidation. Service delivery chains are formed with participants restricting themselves to their core competence viz., manufacture of collectors, tanks etc., distributors and dealers who integrate the plumbing and install the SWH and technicians who provide post-installation service. Ancillary units such as those supplying water softeners and other add-ons take root. Relevant policy measures are enacted to regulate the industry and to encourage and facilitate such consolidation and delineation. Such measures would include creating awareness among banks and financial institutions on the financial, social and environmental benefits of the SWH and ensuring adequate flow of capital for the supply-side and consumer finance for the demand side. For instance, the manufacturers would require capital infusion into plant and machinery while the distributors require working capital facilities.

At this stage, the market constituents who derive the greatest benefit from utilizing the SWH would make the switch first and this could probably be the commercial segment, followed in time by the residential segment. The given demand pattern for heated water, comparatively higher electricity tariff, better affordability and increased borrowing capacity could lead to enhancing demand. Fiscal incentives such as tax-credits and accelerated depreciation benefits are more relevant to this segment, while service provision is more cost-effective. Policy intervention could target greater penetration to this segment to help the industry achieve critical mass.

5.8 Conclusions

Intervention and policy support need to be tailored to suit the stage of the industry's life cycle. Tremendous amount of hand-holding is required at the nascent stages. It is absolutely mandatory, albeit challenging, for the enabling agencies to distance themselves as the industry matures. Allowing the market forces to dictate terms would help make optimum utilization of available resources. Segments deriving maximum benefit from the adoption of SWH (or other RE applications) would voluntarily make the switch. Appropriate financial facilitation is the key to help mainstream the solar thermal technology employed in water heating, or Renewable Energy Technologies in general. Nature, quantum and application of such funding are dictated by the relative maturity of the industry.

References

1. Altener (2002) REACT: renewable energy action, case study # 6—domestic solar water heaters
2. Ardente F, Beccali G, Cellura M, Lo Brano V (2005) Life cycle assessment of a solar thermal collector: sensitivity analysis, energy and environmental balances. Renew Energy 30(2): 109–130
3. Ardente F, Beccali G, Cellura M, Lo Brano V (2005) Life cycle assessment of a solar thermal collector. Renew Energy 30(7):1031–1054
4. Argiriou A, Mirasgedis S (2003) The solar thermal market in Greece–review and perspectives. Renew Sustain Energy Rev 7(5):397–418
5. Audretsch DB, Yamawaki H (1992) Sub-optimal scale plants and compensating factor differentials in US and Japanese manufacturing. In: Audretsch DB, Siegfried JJ (eds) Empirical studies in industrial organization. Kluwer Academic Publishers, The Netherlands
6. Backwell B (2010) Subsidies to be cut for Spain's wind and thermal solar sectors. Recharge News, 5 July 2010. http://www.rechargenews.com/business_area/politics/article219756.ece
7. Bradley R Jr (1997) Renewable energy, not clean, not 'green'. Cato policy analysis 280, Cato Institute, August 27, 1997
8. Chandrasekar B, Kandpal TC (2004) Techno-economic evaluation of domestic solar water heating systems in India. Renew Energy 29(3):319–332
9. Chapman PF (1977) The economics of UK solar energy schemes. Energy Policy 5(4): 334–340
10. Doraswami A (1994) Solar water heating systems in India. Energy for Sustainable Development 1(1):51–57
11. ESTIF (2005) Worldwide capacity of solar thermal energy greatly underestimated. Energy Bulletin, European Solar Thermal Industry Federation. http://www.energybulletin.net/3998.html
12. European Renewable Energy Council (ECRC) (2003) Development of the global solar electricity market. http://www.erec-renewables.org/sources/photovoltaics.htm
13. Greenstein SM, Wade JB (1997) Dynamic modeling of the product life cycle in the commercial mainframe computer market, 1968–1982. NBER Working Paper, W6124
14. Guru S (2002) Renewable Energy Sources In India: Is it Viable? Liberty Institute, New Delhi
15. Hoffman SM (1999) The energy-efficiency/renewable energy industry: attitudes and perspectives concerning electric utility deregulation. Minnesotans for an Energy-Efficient Economy, USA. www.me3.org/projects/dereg/hoffmaneere.pdf

16. Jackson T, Oliver M (2000) The viability of solar photovoltaics. Energy Policy 28:983–988
17. Jimenez V (2004) World sales of solar cells jump 32 percent. Eco-Economy Indicators, Earth Policy Institute, Washington DC, USA
18. Jovanovic B, MacDonald GM (1994) The life cycle of a competitive industry. J Polit Econ 102:322–347
19. Kaldellis JK, El-Samani K, Koronakis P (2005) Feasibility analysis of domestic solar water heating systems in Greece. Renew Energy 30(5):659–682
20. Karagiorgas M, Botzios A, Tsoutsos T (2001) Industrial solar thermal applications in Greece: economic evaluation, quality requirements and case studies. Renew Sustain Energy Rev 5(2):157–173
21. Karlsson C, Nystrom K (2003) Exit and entry over the product life cycle: evidence from the Sweedish manufacturing industry. Small Bus Econ 21(2):135–144
22. Klepper S, Graddy E (1990) Industry evolution and the determinants of market structure: the evolution of new industries and the determinants of market structure. Rand J Econ 21:27–44
23. Klepper S (1996) Entry, exit, growth and innovation over the product life cycle. Am Econ Rev 86:562–583
24. Kolhe M, Kolhe S, Joshi JC (2002) Economic viability of stand-alone solar photovoltaic system in comparison with diesel-powered system for India. Energy Econ 24:155–165
25. Kumar A, Jain SK, Bansal NK (2003) Disseminating energy-efficient technologies: a case study of compact fluorescent lamps (CFLs) in India. Energy Policy 31:259–272
26. Lawley P (2001) Support solar for solar support. Renewable energy generation & certificates trading conference, Sydney, Australia, May 2001
27. McNelis B (1998) PV rural electrification: needs opportunities and perspectives. Presented at the 2nd world conference and exhibition on photovoltaic energy conversion, Vienna, July 1998
28. Milton S, Kaufman S (2005) Solar water heating as a climate protection strategy: the role for carbon finance. Green Markets International, Inc., January 2005
29. Mints P (2010a) FITs: the ecstasy and the agony. Photovoltaics World. http://www.electroiq.com/index/display/photovoltaics-article-display/0722321467/articles/Photovoltaics-World/industry-news/2010/june/fits_-the_ecstasy.html
30. Mints P (2010b) PV industry pricing: the good, the bad, and the confusing. Photovoltaics World. http://www.electroiq.com/index/display/article-display/1113181755/articles/Photovoltaics-World/industry-news/2010/april/pv-industry_pricing.html
31. Mints P (2010c) The PV industry's black swan. Photovoltaics World. http://www.electroiq.com/index/display/article-display/3108489888/articles/Photovoltaics-World/industry-news/2010/march/the-pv_industry_s.html
32. Mishra AK (2009) Sunburnt: Moser Baer's diversification in photovoltaics is scorched by market dynamics. Ratul Puri must take stock. Forbes India magazine, 17 July 2009
33. MNES (2005) Ministry of Non-Conventional Energy Sources, Government of India. http://mnes.nic.in/sw.htm
34. Mueller DC (2003) The corporation: investment, mergers and growth. Routledge, London, p 32
35. Muneer T, Maubleu S, Asif M (2006) Prospects of solar water heating for textile industry in Pakistan. Renew Sustain Energy Rev 10(1):1–23
36. Nyankori JCO, Wabukawo V, Sakyi-Dawson E, Sefa-Dedeh S (2002) Product life cycle model of cowpea based products in Ghana. Clemson University, USA, Working Paper No. WP052402
37. Oliver M, Jackson T (1999) The market for solar photovoltaics. Energy Policy 27:371–385
38. Owen G (2003) GO Solar Company, USA. http://www.solaraccess.com/news/story?storyid=5287
39. Peterson JA (ed) (2000) Micro power–the next electrical era. Worldwatch paper 151, Worldwatch Institute, USA, p 34
40. Polatidis H, Haralambopoulos D (2002) Normative aspects of renewable energy planning; a participatory multi-criteria approach. Working paper, Keele University, School of Politics,

International Relations and the Environment (SPIRE), and International Society for Ecological Economics. (http://www.ecologicaleconomics.org/publica/workpapr.htm)

41. Rossiter D, Wass K (2004) Australia's renewable energy certificate system. Office of the Renewable Energy Regulator, Australia. http://www.orer.gov.au/about/pubs/bonn.pdf

42. Russell J (2010) Record growth in photovoltaic capacity and momentum builds for concentrating solar power. World Watch Institute. http://vitalsigns.worldwatch.org/vs-trend/record-growth-photovoltaic-capacity-and-momentum-builds-concentrating-solar-power

43. Saha PC (2003) Sustainable energy development: a challenge for Asia and the Pacific region in the 21st century. Energy Policy 31:1051–1059

44. Sastry EVR (2003) The photovoltaic market in India. IEA – PVPS Conference, Osaka, Japan, May 2003

45. Seetha (2003) People Power: Market For Electricity From Below. Liberty Institute, New Delhi, India

46. Shivakumar AR (1996) A value addition to a modern house design: an integrated domestic solar water heater. Energy for Sustainable Development 3(1):54–58

47. SolarAccess.com (2002) Consolidation Planned in Solar PV Industry. http://www.solaraccess.com/news/story?storyid=1415

48. Soya Choupal (2011) ITC: International Business Division : www.soyachoupal.com

49. Srinivasan S (2005) Segmentation of the Indian photovoltaic market. Renew Sustain Energy Rev 9:215–227

50. Stromsta K-E (2010a) Q-Cells changes direction, becomes module maker. Recharge News, March 24 2010. http://www.rechargenews.com/energy/solar/article209387.ece

51. Stromsta K-E (2010b) Tata BP Solar starts up new production line in Bangalore. Recharge News, April 28 2010. http://www.rechargenews.com/energy/solar/article213306.ece

52. TERI (2001) Survey of renewable energy in India. Tata Energy Research Institute, New Delhi

53. Turkenburg WC (2000) Current status and potential future costs of renewable energy technologies, World Energy Assessment report, prepared by UNDP, UNDESA and the World Energy Council. United Nations, New York

54. Wang Z (2003) New product diffusion and industry life cycle. Department of Economics, University of Chicago Illinois, USA

55. Werker C (2000) Market pewrformance & competition: a product life cycle model. Discussion Paper, Greifswald University, Germany

56. Wesoff E (2010) Global solar markets: end of the gold rush era. Greentech Media. http://www.greentechmedia.com/articles/read/global-solar-markets-end-of-the-gold-rush-era

57. Winkler H (2005) Renewable energy policy in South Africa: policy options for renewable energy. Energy Policy 33:27–38

58. Wohlgemuth N, Madlener R (2000) Financial support of renewable energy systems, investment versus operating cost subsidies. Conference Proceedings—Towards an Integrated European Energy Market, Norway, 2000

59. Xiao C, Luo H, Tang R, Zhong H (2004) Solar thermal utilization in China. Renew Energy 29(9):1549–1556

Chapter 6
Accounting for the Environmental Externality

> ..*sustainability means running the global environment - Earth Inc. - like a corporation: with depreciation, amortization and maintenance accounts.*
>
> Maurice Strong
> First Executive Director of the UNEP

Renewable Energy programs are often sought to be justified on the basis of the private benefits and costs accruing to the individual households, in terms of providing improved lighting, superior cooking fuel, improved indoor air quality and the like. Benchmarking electric power or other services against their respective closest substitutes, namely power from coal fired plants or services provided by burning fossil fuels is incomplete if the differences in environmental impact are not taken into account. This chapter discusses the positive environmental externality accruing from domestic RE programs, to demonstrate that economic surpluses from domestic programs are realized beyond narrowly defined project boundaries. Employing biogas programs to illustrate, it is shown that the economic value addition from the consumptive use of the biogas for cooking, and the non-consumptive and indirect value derived from the biogas plant, viz., providing feedstock for other processes and other such benefits as greenhouse gas mitigation (positive externalities) need to be accounted for. The process approach adopted herein enables an integrated view of the value chain and consequently, a mechanism to reallocate costs and to distribute such surpluses.

Biogas digesters have come to symbolize access to modern energy services in rural areas and are slated to considerably improve health and sanitation, and to yield significant socio-economic and environmental benefits.[1] The gas released

---

[1] This chapter is largely based on Srinivasan, Sunderasan "Positive Externalities of Domestic Biogas Initiatives: Implications for Financing", Renewable and Sustainable Energy Reviews 12, (2008), pp. 1476–1484, Elsevier Limited.

S. Sunderasan, *Rational Exuberance for Renewable Energy*,
Green Energy and Technology, DOI: 10.1007/978-0-85729-212-4_6,
© Springer-Verlag London Limited 2011

from the anaerobic digestion of animal residues, when used as a cooking fuel, provides for superior combustion and displaces dirtier and less efficient cooking fuels viz., firewood. By reducing indoor smoke and consequent ocular and respiratory infections, biogas digesters contribute to improved health and to proportional reductions in medical expenditure. The surplus gas could be put to use on other applications, viz., water and space heating, [2], and could potentially replace conventional alternatives. Batzias et al. [5] have evaluated energy and biogas potential of livestock residues and have presented a GIS based biomass resource assessment application to explore the possibility of upgrading the biogas and transporting it through a nationwide pipeline network. Plant owners have the option of linking sanitary toilets to the digesters thus improving hygiene, though this proposition often encounters social reluctance.

Promotional measures, interest subsidies and cost buy-downs encourage construction of biogas plants (or without loss of generality, other equipment such as wind turbines, solar PV systems etc.) but do not necessarily, ensure continued operation, which is essential for the environmental benefits and consequent cash-flows to accrue. The convenience of cleaner fuel with superior combustion and one providing for improved indoor air quality is, by itself, not adequate motivation to alter life styles, especially in the face of handy availability of alternatives such as fuel wood. Further, inadequately informed farmers fail to distinguish between raw animal residues and the digestate, thereby wasting the superior manorial value of the latter. Microfinance schemes for the biogas program in developing countries, thus, have to be designed to *facilitate* the construction of new plants, and more importantly, provide *incentives* for sustained operation. Such facilitation could take the form of working capital facilities for vendors, awareness campaigns and quality assurance operations. Incentives could emanate from creating markets for raw dung, the gas itself, or the digested slurry priced as above. The microfinance institution (MFI) could diversify its exposure and ensure commercially viable returns on its portfolio, by balancing between low interest credit to individual farmers and commercial terms and equity-like instruments for small business ventures. Multilateral and donor agencies could invest their contributions in the facilitation phase and then pave the way for market mechanisms to sustain the momentum.

6.1 Value-Addition from Biogas

The benefits of biogas digesters and its products are well known. Construction of the biogas plants generates employment opportunities by itself. Further, women save several hours a day on firewood collection, cooking and cleaning soot deposits off cooking pots. Plant owners, thus, accrue additional benefits from diverting the saved time to productive activities. The slurry produced from the digestion process has superior manorial value as compared to the raw animal waste and finds several applications, [14]. As an organic manure, it replaces chemical fertilizers, enhances farm yields and when certified as organic produce, commands premium revenues, especially when exported to developed country markets. The biogas plant effluent is also used as an organic fertilizer in fish polyculture, substantially growing fish yields [4, 23], and for various other purposes, including, as an absorbent for removal of lead from manufacturing industry waste water [19].

Several *indirect* benefits flow from large volume installation of family sized biogas digesters. The most notable, of course is the mitigation of methane, a potent greenhouse gas (GHG) released from the open decomposition of animal wastes and the avoided carbon-di-oxide release from burning firewood. The reduced use of firewood retards deforestation and the use of organic manures improves the physical, chemical and biological properties of soils while also providing for a vital input to the organic farm trade. Soil regeneration is aided by the farmers resorting to stall feeding their cattle which enables them to capture manure conveniently. As a result, public lands are spared from overgrazing, and are more sanitary.

Domestic biogas programs (or other rural RE initiatives) are frequently justified on the basis of the *private* benefits and costs accruing to the individual households, in terms of providing a superior cooking fuel, improved indoor air quality and saving of time spent on collecting firewood. For instance, Bala and Hossain [3], evaluate the economics of biogas digesters in Bangladesh in terms of firewood and fertilizer values. Tampier [22] recounts examples across applications and from around the world, and concludes that "distributed energy utilities" could achieve greater market penetration with greater awareness and appropriate consumer financing. Policy makers need to realize that the economic surpluses from domestic biogas programs are realized beyond such narrowly defined project boundaries. Such positive externalities imply that the total benefits accruing from the installation of biogas plants exceed the benefits to the individual who receives the service. Society at large, is perhaps, likely to benefit more than the individual recipient availing of the biogas digester, [8].

Lending a time dimension, Pehnt [20] investigates a dynamic approach to life cycle assessment (LCA) of Renewable Energy Technologies and observes that prospective product and process development would progressively improve the environmental characteristics of such applications. In this chapter, we evaluate ripple effects and lay out a framework to estimate the economic value addition from the consumptive use of the biogas for cooking, its non-consumptive (existence) value from disposing raw animal dung, and the *indirect* value derived from the biogas plant providing feedstock for other processes and other such benefits as greenhouse gas mitigation.

6.2 Relevant Experiences and Matters of Concern

Iqbal Quadir [12] compares technology to an "invisible leg" capable of moving the economy from one state to the other by providing connectivity, wider access to markets and by preventing waste, resulting in GDP growth rates far greater than those achieved by repeated infusions of foreign aid. Van Groenendaal [24] has computed the net present value of future benefits in various manufacturing sub-sectors and has assessed the demand for a new fuel (natural gas) by forecasting growth in gross value added. Using Vietnam as a case illustration, Dang et al. [10] demonstrate the integration of mitigation and adaptation strategies that can provide additional benefits to social welfare in addition to climate change mitigation. Shi and Gill [21], have found that the environmental soundness of technology viz., biogas digesters alone is insufficient to entice farmers. They list the limited availability of information, risk aversion and high transaction costs as major barriers to the adoption of alternative practices. An objective assessment of ben-efits and costs and an optimal redistribution thereof, as proposed by this chapter, should help overcome cost and risk related barriers.

An analysis of energy balances from a life cycle perspective of large scale biogas plants operating in Swedish conditions, indicates that the net energy output turns negative when transport distances of feedstock manure are large, [6]. In rural India and other developing countries, household biogas plants are sited in close proximity to both the cattle sheds and to cattle farmers' residences where the gas is slated to be consumed, and hence to yield a positive net energy output.

On analyzing the phase out of leaded gasoline, Hilton [15] concludes that while high income does not guarantee pollution abatement, poverty does not prevent it either, and that "poor nations can and should plan to reduce pollution before they become rich." Lichtman [16] highlights the importance of basing technology deployment programs on an in-depth "understanding of rural resource flows and the local political economy." He suggests that the growth of local organizations would ensure that the "people focus" of development assistance is retained. Malhotra et al. [17] discuss the role of women in household energy management and present a participatory process illustrating their involvement in rural energy decisions.

Allen and Loomis [1] derive non use and non-consumptive use values for wildlife and arrive at willingness-to-pay estimates, potentially leading to policy decisions that are based on an objective understanding of benefits and costs from forest projects. Brennan [7] analyses "green preferences" and proffers that if sufficient numbers of consumers willingly switch to low-emission technologies, tax or permit policies become less necessary or stringent. Ghosh [13] discusses the possible short-run reduction in farm yields on making the switch from chemical fertilizers to organic manure and recommends that the negative effects could be mitigated through appropriately structured compensation for the farmers and by promoting a dynamic manure market. This chapter, and this book in general, envisages that "green" preferences would be made voluntarily, if substantiated by economic incentives and a fair redistribution of benefits and costs, such as through creating a dynamic market for biogas digestate and that compensation required to alter lifestyles would take the form of increased total revenues from organic farm produce.

6.3 Function Modeling

The Integrated Definition Method (IDEFØ) is designed "to model decisions, actions and activities of an organization or system"[2] and is especially useful for functional analyses and for definition of the inputs, outputs, resources employed and the constraints the function is subject to. This discussion uses the process mapping technique to bring out the differences between the pre and post biogas scenarios, Figs. 6.1 and 6.2 respectively, and to illustrate the link between a 'first world' consumer's quest for clean air and healthy food, and a biogas plant in a developing country. The map in Fig. 6.3, further summarizes the private, local and global benefits and beneficiaries from the large scale installation of biogas digesters.

6.3.1 Private Benefits and Costs from Digester Installation

The cash flow streams associated with the ownership and operation of the digester are as follows: Cash out-flows on plant acquisition and procurement of the raw dung (or foregone revenue from sale of raw dung) and inflows from avoided firewood procurement costs and from the sale of digested slurry to a downstream process. The improved health of residents and avoided medical expenditure enhances plant owner liquidity. The justification for the investment would be "reasonable" payback periods, approximately overlapping with the tenure of the loans on the plant.

6.3.2 Benefits of Improved Health

Health has both direct micro and indirect macro effects on a country's economy; direct through the impact of ill health on current productivity, and indirect via the size and quality of the labor force as determined by such factors as mortality, fertility and intellectual capacity. Studies examining the effect of ill-health at a household level tend to underestimate the full economic impact of ill-health and the benefits from improved health as supply of unskilled and semi-skilled labor tends to be rather inelastic owing to financial compulsions. However, society values the individual health benefits from such applications, for, individuals tend to underestimate the costs imposed on the rest of society by communicable diseases. Mills and Shillcutt [18] have estimated the benefit/cost ratios for Malaria control. The have computed that providing 60 million additional children with insecticide-treated mosquito nets with an investment of $1.77 billion would yield a benefit of over $18 billion i.e., a factor of 10.

---

[2] http://www.idef.com/IDEF0.html.

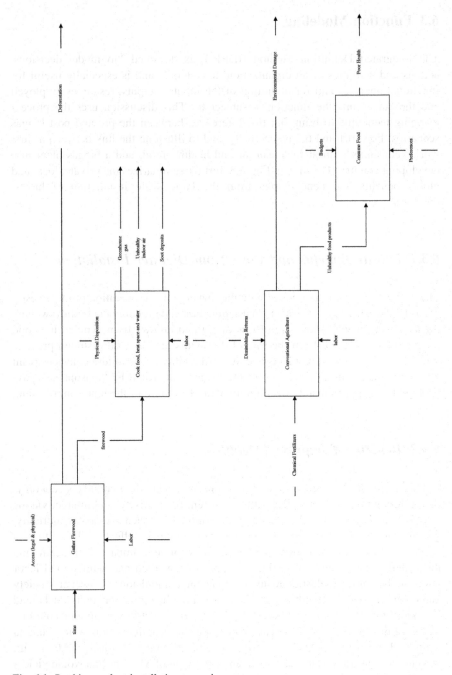

Fig. 6.1 Pre biogas plant installation scenario

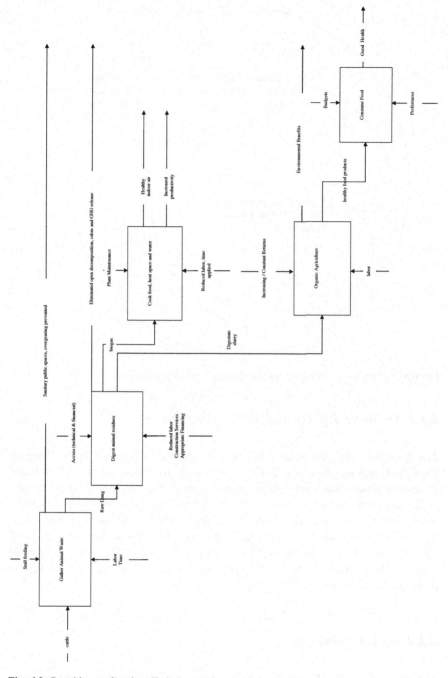

Fig. 6.2 Post biogas plant installation scenario

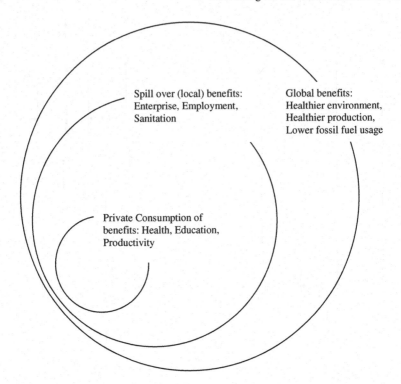

Fig. 6.3 Private, local and global benefits derived from biogas digesters

6.3.3 Spillover Effects and Value Creation

The firewood collection process is now redundant as is the need for washing of blackened cooking pots, thus, freeing up resources for productive use in the order of several person-hours per day per plant installed. Further, improved health liberates time spent nursing the sick. Surplus time now available could be monetized at prevailing wage rates. The deforestation thus avoided, amounting to about two tons of firewood per average biogas-household creates non-consumptive and indirect value from sequestration. The replacement of kerosene lighting with superior illumination provides for longer study hours for children, with expected positive returns in the medium-term.

6.3.4 Social Value

The construction and financing of biogas digesters creates employment opportunity by itself, contributing significantly to the rural economy. Stall feeding of cattle to enable convenient collection of dung and linking of sanitary toilets to the digester leads to 'unintended consequences' of avoided overgrazing and more sanitary public spaces.

6.3.5 Enterprise Value Creation

The distinguishing aspect of the process approach discussed herein is, portraying the digestate slurry produced from the digestion process as a marketable product, yielding revenue streams to the plant owner household and feeding into a downstream process. The biogas plant is thus rendered a self-liquidating asset, paid for by these subsequent processes.

The market for organic produce is large and growing. Taking note of the rising rates of obesity and compounding problems, viz., diabetes, heart disease, stroke, cancer and birth defects from pesticides, the US Department of Agriculture estimates that healthier diets alone could obviate US$71 billion in annual medical expenses. Concomitant with the increasing awareness that sustainable agriculture works best for both people and nature, the demand for organic products has grown at about 20% per annum with the US alone consuming about $12.8 billion worth of such food and beverage in the year 2005 [9].

As for pricing, it is estimated that [11] organically grown teas command a premium of 30–40% over conventionally produced tea and is rendered remunerative despite increased labor intensity and lower productivity in the short run. A portion of this surplus could be invested in securing supplies of superior organic manures emanating from biogas digesters.

The local and regional economy gains from the additional inflows from the sale of organic products and from the cash liberated from medical expenditure. More importantly, the infusion of cash into the local economy gets multiplied by the secondary consumption expenditure it triggers. While the savings provide the MFIs with funds for on-lending, each additional unit of currency entering the local economy from the outside results in more than one unit of additional output and consumption, overall (expenditure multiplier in action).

6.3.6 Global Benefits

The immediately recognized global benefits pertain to the mitigation of green house gas emissions; diminished grazing and retarded deforestation yield sequestration benefits, as well. A secondary impact is the reduction in demand for kerosene and other distillates derived from finitely available fossil fuels. The plants help secure a steady and increasing supply of labor-intensive organic produce, at which the developing countries clearly enjoy a comparative advantage.

Thus, the case for positive externalities [8] necessitating transfer payments is quite strong, as the total marginal benefit accruing from the biogas plant exceeds the private marginal benefit—possibly leading the individual to undervalue the facility on offer.

6.4 Conclusion: Assignment of Benefits and Beneficiaries

While the benefits from plant installation and usage flow from the left to the right of the process map, (Fig. 6.2), the compensation for the same flows in the opposite direction. The economic agents deriving the health benefits from the consumption of organic produce pay a premium over conventional agricultural produce which compensates the organic farms and ultimately trickles down to the biogas-household in the form of slurry revenues. A second stream, normally in the form of development assistance or of procurement of 'carbon credits', offers a relatively less expensive mitigation initiative for a population charged with the responsibility of mitigating carbon emissions, (Fig. 6.3).

Sustainable microfinance programs are required to be consistent with cost recovery and simultaneously to operate on the strength of social collateral and flexibility. Programs enabling the installation of biogas digesters could, thus, de-risk their portfolio by recovering (or securing) a part of their receivables from the organic farmers or cooperatives, who generally tend to be larger and more solvent than small farmers and individual cattle owners. The split in payment responsibility could reflect the (present discounted) value derived from the use of biogas as a cooking fuel and the slurry as an input to the farm. The compensation for global benefits (positive externalities accruing beyond local boundaries) could then be used to foster a conducive environment by building institutions and market practices viz., awareness creation, quality assurance etc., benefiting all the agents concerned. The next chapter focuses on the role played by micro-finance institutions in enhancing the penetration or RE technologies in rural areas and the incentives that could be provided for sustained operation.

References

1. Allen BP, Loomis JB (2006) Deriving values for the ecological support function of wildlife: an indirect valuation approach. Ecol Econ 56:49–57
2. Axaopoulos P, Panagakis P (2003) Energy and economic analysis of biogas heated livestock buildings. Biomass and Bioenergy 24(3):239–248
3. Bala BK, Hossain MM (1992) Economics of biogas digesters in Bangladesh. Energy 17(10):939–944
4. Balasubramanian PR, Kasturi BR (1994) Biogas-plant effluent as an organic fertilizer in fish polyculture. Bioresour Technol 50(3):189–192
5. Batzias FA, Sidiras DK, Spyrou EK (2005) Evaluating livestock manures for biogas production: a GIS based method. Renew Energy 30(8):1161–1176
6. Berglund M, Borjesson P (2006) Assessment of energy performance in the life-cycle of biogas production. Biomass and Bioenergy 30(3):254–266

7. Brennan TJ (2006) Green preferences as regulatory policy instrument. Ecol Econ 56:144–154
8. Clarke GRG, Wallsten SJ (2003) Universal service: empirical evidence on the provision of infrastructure services to rural and poor urban consumers. In: Brook PJ, Irwin TC (eds) Infrastructure for poor people. The World Bank, Washington DC
9. Cummings CH (2006) Ripe for change: agriculture's tipping point. World Watch Magazine, July/August 2006, World Watch Institute. www.worldwatch.org
10. Dang HH, Michaelowa A, Tuan DD (2003) Synergy of adaptation and mitigation strategies in the context of sustainable development: the case of Vietnam. Climate Policy 3(Suppl 1): S81–S96
11. DTRDC (2003) Organic tea cultivation. The Darjeeling Tea Research and Development Center, Government of India. http://www.dtrdc.org/organic.htm
12. Economist (2006) Power to the people. The Economist, 9 Mar 2006
13. Ghosh N (2004) Reducing dependence on chemical fertilizers and its financial implications for farmers in India. Ecol Econ 49(2):149–162
14. Gurung JB (1997) Review of literature on effect of slurry use on crop production. Biogas Support Program, Nepal, June 1997
15. Hilton FG (2006) Poverty and pollution abatement: evidence from lead phase-out. Ecol Econ 56:125–131
16. Lichtman R (1987) Toward the diffusion of rural energy technologies: some lessons from the Indian biogas program. World Development 15(3):347–374
17. Malhotra P, Neudoerffer CR, Dutta S (2004) A participatory process for designing cooking energy programmes with women. Biomass and Bioenergy 26(2):147–169
18. Mills A, Shillcutt S (2004) Communicable diseases: summary of Copenhagen consensus challenge paper. May 2004. www.copenhagenconsensus.com
19. Namasivayam C, Yamuna RT (1995) Waste biogas residual slurry as an adsorbent for the removal of Pb(II) from aqueous solution and radiator manufacturing industry wastewater. Bioresour Technol 52:125–131
20. Pehnt M (2006) Dynamic life cycle assessment (LCA) of renewable energy technologies. Renewable Energy 31:55–71
21. Shi T, Gill R (2005) Developing effective policies for the sustainable development of ecological agriculture in China: the case study of jinshan county with a systems dynamics model. Ecol Econ 53(2):223–246
22. Tampier M (2006) Distributed energy utilities—it is all about financing, renewable energy access, 22 May 2006, http://www.renewableenergyaccess.com/rea/news/story?id=44964
23. Tripathi SD, Karma B (2001) Biogas slurry in fish culture. In: Integrated agriculture–aquaculture: a primer, Fisheries Technical paper no. 407, Food and Agriculture Organization
24. Van Groenendaal WJH (1995) Assessing demand when introducing a new fuel: natural gas on Java. Energy Econ 17(2):147–161

Chapter 7
Microfinance: Taking RE to the 'Former Poor'

"My vision for the future: to make credit a human right.."
Muhammad Yunus
Founder, Grameen Bank, Bangladesh and Nobel laureate

Through most of the length of this text, we have discussed the significance of economic variables such as prices, fiscal incentive structures and of the importance of tailoring financing structures to suit the product and the stage in its life-cycle. In this chapter we discuss the role played by micro-finance institutions in enhancing the uptake of (primarily standalone) RE technologies. Micro-banking facilities have helped large numbers of developing country nationals claw their way out of abject poverty, specifically by supporting the establishment and growth of microenterprises. And yet, the microfinance movement has grown on the back of passive replication and the movement now needs to be revitalized with new product offerings and innovative service delivery. Domestic RE systems may not inherently be income generating, and returns on such investments accrue, primarily, from cost avoidance. By definition, and often by explicit regulatory limitation, domestic RE applications do not qualify for micro-funding. 'Quality of life' investments such as RE systems, funded through non-self-liquidating loans, reflect borrower maturity and simultaneously contribute to MFI sustainability. Larger loan amounts and longer tenures reduce credit appraisal and service delivery costs per dollar lent. By promoting the consumption of RE systems, MFIs would be contributing to environmental conservation, as well. Innovative institutional structures need to evolve to facilitate the progression of the MFIs into more formal outfits, capable of accessing refinance and capital from banks and commercial investors, and of providing larger loan sizes for longer tenures.

Dr. Muhammad Yunus: "After the year 2005, Grameen, the bank which is known all over the world as a 'bank of the poor', should acquire a new identity; it should be known as the 'bank of the former poor'. This is our challenge", Presentation to the Director General of UNESCO, 1994.

S. Sunderasan, *Rational Exuberance for Renewable Energy*,
Green Energy and Technology, DOI: 10.1007/978-0-85729-212-4_7,
© Springer-Verlag London Limited 2011

The strong positive correlation between the lack of financial access and low incomes is well established [5] and credit is widely perceived as a model instrument to end poverty, [16]. Such credit offers "hope to many poor people of improving their own situations through their own efforts", [17]. Review and evaluation assessments of microfinance programs have reported economic prosperity of borrowers, confirmed by reduced vulnerability, increased incomes from diversified and secondary sources, increased savings, and 'consumption smoothing', [25]. Evidence from Bangladesh suggests that the provision of microcredit leads to "asset creation, employment generation, economic security and empowerment of the poor", especially the women, [12]. Ruben and Clercx [34] have found that access to rural finance could also reinforce food security and facilitate income diversification, while leading to ecological conservation. Micro-banking, essentially, replaces complicated credit-evaluation and collateral requirements with lower-cost procedures based on trust and solidarity.[1]

Microfinance (MF) is viewed as "socially responsible", justified by the notion of *fairness* towards the poor, or of financial *equity* attained by profitably supporting their enterprise. "The delivery of microcredit takes place within an environment where the different financial market players face their own set of constraints in supplying credit to small-scale borrowers", [8]. In industry parlance, Grameen replication refers to 'reproducing the essential features' of the Bangladesh-Grameen-approach in different settings across the world. Most replicators, for instance, commence with small and progressively growing loan sizes, and lend to groups of poor women while insisting on weekly repayments, [27]. Yet, practical innovation and change are essential while retaining focus on the poorest women and on ensuring discipline and high recovery rates, [39, 51]. Progressive evolution and optimization are integral to the development of the MF industry, and quality and customization are sometimes overlooked in the rush to expedite replication, [50].

The success of the micro insurance schemes in India, Uganda and Bangladesh [10] and the Grameen microleasing initiative serve to illustrate that the poor have diverse credit needs and that the Microfinance Institutions (MFI) are required to provide responsive and adaptable products, [44]. After demonstrating, first, that the poor are indeed bankable, and then signaling their demand for diverse offerings, it is now time to offer a portfolio of products and value-added services, to strike a balance between the needs of the poor on the one hand, and the need for institutional financial sustainability on the other, [3, 29]. In fact, the lack of product innovation and the dissatisfaction with current offerings, often lead to higher MFI client dropout rates.

---

[1] This chapter is largely based on Srinivasan, Sunderasan "Microfinance for Renewable Energy: Financing the 'former poor'", World Review of Entrepreneurship, Management and Sustainable Development, Vol. 3, No. 1, 2007 Inder Science, UK, pp 79–89.

7.1 Rural (Stand-Alone) Renewable Energy Applications and Defined Impediments

Off-grid Renewable Energy (RE) finds application in cooking, lighting, process motive power, water pumping, heating and cooling and in filling other small electrical needs. Improved cook stoves consume a fraction of the biomass while also dramatically improving indoor air quality, [26]. As discussed in Chap. 6, the installation of biogas digesters reduces workload, typically, for women and girls, saves on kerosene and fuel wood and hence retards deforestation, reduces greenhouse gas emissions, improves indoor air quality, improves sanitation and hygiene, generates slurry used as organic fertilizer or fish food. Solar home systems (SHS) and family hydro systems provide lighting and power radios, televisions or electric fans thereby replacing conventional alternatives viz., candles, kerosene lamps, storage batteries and diesel generators while helping improving indoor air quality and health, simultaneously reducing greenhouse gas emissions. Solar lanterns find multiple applications, owing to their portability. Solar water heating systems provide hot water to meet various domestic and commercial requirements.

The assembly, sale and service of RE systems offer employment opportunities by themselves, [42]. We are, further, required to distinguish between commercial (micro-enterprise) and domestic applications of these technologies. Hospitals, rural clinics and dispensaries need power for lighting and for refrigerators containing vaccines and medicines. Solar PV systems at small shops, solar lanterns for poultry farms, or biogas for tea stalls or small restaurants would be considered 'productive' or income generating and hence, would automatically qualify for microfinance loans. Domestic application of these technologies, on the other hand, provides physical comfort, avoids costs by substituting traditional alternatives, improves indoor air quality and extends study hours for children. And yet, most of these appliances or applications may not qualify for microfinance loans, as they are not considered 'productive' or revenue generating. This is in many instances, explicitly restricted by law.

The Moroccan Law on Microfinance, for instance, purports to regulate the industry and spells out several incentives, such as liberating microfinance institutions (MFI) from interest rate caps normally imposed on banks and finance companies, permitting the charge of fees and exempting MFIs from value-added taxes for five years, [11]. Notably, in keeping with the Grameen concept of supporting income-generating activities, the 1999 Law explicitly restricts microcredit loans to 'productive' activities to help the people "succeed economically". Bolivian Law defines microcredit as "loans granted to a borrower or group of borrowers that uses a solidarity guarantee for financing *productive activities.*"[2] The Rural Microfinance Development Center (RMDC) [33], Nepal provides wholesale credit and institutional support to MFIs while restricting the ultimate

---

[2] http://microfinancegateway.org/resource_centers/reg_sup/micro_reg/country/5/.

borrowing to income-generating activities. India defines micro-banking as a service which provides small amounts of thrift, credit and other financial services and products to the poor for *income generation*.[3] The Microfinance Development Strategy formulated by the Asian Development Bank (ADB) [1] aims to ensure institutional financial services for small businesses. In some of these cases, concerted efforts have been invested to amend the provisions to enable lending for small scale domestic RE systems.

7.2 Microfinance and Non-Self-Liquidating Loans

The exclusion of consumption loans is by no means universal. South African Law,[4] for example, specifies loan tenures and ceilings on loan sizes but does not define asset class(es) that qualify for microfunding. The 2002 Kyrgyz Republic[5] Law on Microfinance Organizations permits the MFI to define its own scope and limitations in accordance with its charter. Kenya[6] explicitly provides for loans to micro or small enterprises *and low-income households*. The Central Bank of Philippines[7] offers a broader definition of microfinance in terms of loans to the poor and low-income houses to raise income levels *and living standards*.

Precedents from different parts of the world could be cited, where similar 'non-productive' or non-self-liquidating loans have been made available. One of the high priority items for micro-entrepreneurs, especially women members, is investment in children's education [13]—in getting children to school and keeping them there longer—and consequently, several MFIs have tailored savings and credit programs to meet schooling expenses, [25]. Conventional education is clearly a 'long-lead' investment with stochastic returns and is not self-liquidating in the short run. MFIs in Africa and a few other parts of the world, often integrate non-credit add-on services such as health education and HIV/AIDS prevention and improved nutrition [15], recognizing the challenging circumstances under which the woman borrowers operate, and acknowledging that growth and sustainability of the MFI are inextricably intertwined with borrower well-being.

Despite costing substantially higher than the prime interest rate, the housing microfinance loan for minor or major repairs or extensions is a sought after product, since it comes packaged with technical assistance and vastly simplified guarantee and collateral requirements. In this context, the MFIs are ideally placed to offer smaller loans of the order of USD250 – USD500 of short tenures while commercial banks are better equipped to offer relatively larger loans for tenures

---

[3] http://microfinancegateway.org/resource_centers/reg_sup/micro_reg/country/19/.

[4] http://microfinancegateway.org/resource_centers/reg_sup/micro_reg/country/40/.

[5] http://microfinancegateway.org/resource_centers/reg_sup/micro_reg/country/26/.

[6] http://microfinancegateway.org/resource_centers/reg_sup/micro_reg/country/24/.

[7] http://microfinancegateway.org/resource_centers/reg_sup/micro_reg/country/37/.

ranging from 5 to 10 years or more. The loan product helps diversify MFI exposure, reduces monotony and retains clients, and could possibly help mobilize low cost refinance from housing specific funds, [19]. Housing and infrastructure-related loans constitute almost half of all loans issued by SEWA Bank, Gujarat, India, which views this distinction between productive and non-productive home-improvement loans as "artificial", since many self-employed women such as weavers and tailors work out of their homes, [6].

Household water supply and sanitation schemes require up-front investments and the demand for such conveniences (in truth 'necessities') is large and growing in the developing world. Darren Saywell and Fonseca [37] compare sanitation projects across countries and examine the micro-credit mechanisms employed to support such initiatives. Varley [45] argues that such investments are not only attractive but also viable as they generate personal economic benefits and hence consumers are *willing and able* to pay for them. The same has also been under-written by the Copenhagen Consensus[8] initiative, seeking to prioritize global challenges. The time gained from avoided water conveyance and better health from improved sanitation translate into better and sustained work performance, time saved on nursing sick children, and regular attendance at school for children. Rijsberman [32] estimates that an investment of approximately USD11.1 billion in community managed small-scale water supply and sanitation in the developing world would yield a positive Net Present Value (NPV) of USD400 b in saved time and consequent productivity.

7.3 Microfinance for Micro-Infrastructure

As discussed above, renewable energy (RE) systems such as solar home systems and biogas digesters, when applied for domestic consumption do not, *necessarily*, yield incomes directly. However, energy is integral to development and poverty alleviation, [9] and the lack of access to credit is a direct and very obvious barrier to RE technology deployment, [48]. Besides improving indoor air quality and increasing the free time available, [14], such systems avoid costs on expensive traditional alternatives such as candles, dry cells, kerosene, diesel. Improved health augments productivity and reduces medical expenses. *Avoided cost, in this context, needs to be treated as notional income.*

The environmental benefits from cleaner technologies and the avoided clean-up costs, and the revenue streams generated from longer working hours and from better education, are certain to yield benefits which are a huge multiple of the investments made. Even if such additional incomes are not taken into account, there is a need to redefine microfinance, and to convince the MFIs and the policy making authorities that returns from domestic RET utilization accrue from cost avoidance as opposed

---

[8] www.copenhagenconsensus.com.

to revenue generation. Characterizations need to be rephrased to include or at the least not preclude the utilization of microloans for domestic RE applications. Countries where such laws are yet to be framed would do well to adopt a broader definition for the scope and conduct of microfinance activities.

7.4 Larger Loan Sizes for RE Systems

Credit for domestic RE systems falls on the fringes of microfinance, from yet another perspective: that of loan size. Most programs start with small loan sizes and progressively nurture their borrowers into accessing and repaying larger loans. A domestic RE application requiring a loan of, say, USD400 would therefore qualify after about four rounds of borrowing, or about five years into microcredit membership. Several programs explicitly impose a ceiling (in the range of USD300) on maximum loan sizes, [30]. Larger loan sizes reflect borrower maturity and MFI mission accomplishment, and if they are to stay in business, MFIs need to be flexible enough to evolve with their clients.

At the least, such ceilings need to provide for prevalent consumer inflation in the given context. Further, whether a loan is "large" or "small" depends on borrower and lender ranking of aspects of the loan beyond the disbursed dollar amounts and consequently, measures of loan size comprise multiple dimensions, viz., tenure of the loan, number and frequency of repayments, average balance etc., [38]. A longer term to maturity, for instance, is a larger loan compared to a shorter term of the same dollar lending, and provides larger interest income to the lender from a single evaluation and disbursement. On the other side, a longer loan is an indicator of borrower maturity and increased creditworthiness. Dollar-years of borrowed resources accounts for time and "measures the purchasing power provided by the loan and the time through which the borrower controls this purchasing power."

Loan growth is vital for financial sustainability but also serves as a proxy for positive impact. Financial sustainability of microfinance programs presumes a declining cost per dollar loaned, as members' loan sizes grow. Systematically conducted loan product development, keeping both, the customer needs and the institutional strengths and constraints in mind would reduce drop-outs, attract fresh clients, and hence, contribute to the long-run sustainability of the MFI, [49]. Microfinance programs are designed to reduce poverty and increase income with the ultimate objective of enhancing the quality of life of their clientele. The ability of the clients to borrow and repay progressively larger amounts is evidence, on the one hand for client stability and viability of the program, and poverty alleviation of the target segment, on the other, [30].

MFIs impose such limitations to stay within the borrowers' capacity to repay, while also reaching out to as many borrowers as possible with available loan funds. Some authors have actually argued against progressively rising loan sizes, [22]. But, enforcement problems induce restrictions on loan size while rising 'per-borrower' costs necessitate larger loan sizes. The borrowers are slated to have established and reliable sources of income and track records for repayment by the time they seek out 'quality of life' loans. Further, innovative attempts have been made to combine income-generating opportunities, simultaneous with the sale of energy systems and services, [46].

As for refinancing their portfolios, professionally managed and transparent MFIs would be able to tap public deposits, local and international creditors and investors, as well as sector-specific funds such as those for housing or energy. The Deutsche Bank Microcredit Development Fund, for instance, provides catalytic funds and facilitates the formation of durable relationships between MFIs and local financial institutions.[9] Lowering transaction costs and the creation of a secondary market for the investment instruments would help attract mainstream investors, [20].

7.5 Collateral Substitutes and Lowering the Costs of Intermediation

Microcredit is expensive [23] and a majority of the MFIs are unlikely to attain financial self-sufficiency, [43]. The cost of an efficiently delivered microloan could be as high as 51% of the outstanding balance and yet women (who constitute a large proportion of the borrowers) around the world have willingly borne the cost by running microenterprises not only to reduce their own poverty but also to provide a better future for their children, [18]. While such interest rates sound usurious, it has been shown repeatedly that the problem for low-income borrowers is access to credit rather than the absolute interest they must pay, as measured in currency units.

Microfinance in India and a few other countries has evolved along two distinct channels: the 'grameen class' independent MFIs and the ('solidarity') 'self-help group (SHG) – Bank linkage' model. The latter configuration is a quest to blend the safety and reliability of formal finance with the convenience and flexibility attributed to informal finance. Banks, typically, commence with a loan amounting to four times the group's savings for onlending within the group on flexible repayment schedules. The group members, in this scenario, thus save, borrow and repay collectively. The SHG is a de facto MFI availing refinance from a rural or commercial bank. The SHG-bank model in India is touted as the most promising channel for mass-outreach banking, with the number of groups slated to have crossed the one million by 2008, [4].

The solidarity group arrangement has several social and economic returns. The community members often have better information than banks and hence peer screening enhances accountability while improving portfolio quality, consequently reducing transaction costs for banks. The urge to preserve social ties ensures disciplined loan servicing, as limited wealth, limited property rights and poorly

---

[9] http://www.community.db.com/downloads/dbmicro_dev_fund04.pdf.

functioning legal systems combine to render the seizure of collateral, near impossible. While such 'social sanction' is one substitute for physical collateral, *the potential denial of future credit is an equally robust alternative*, [7]. Lending to groups of poor clients, therefore, calls for an atypical outlook from traditional bankers, who, consequently, need to be "conditioned" by match-makers viz., NGOs and consultants, [21].

An extension of the SHG-Bank model to larger loan sizes for energy micro-infrastructure has a number of advantages. Despite having remained with the program for almost the same tenure, it is unlikely that group members decide to acquire RE systems, all at the same time. Consequently, the group loan size and the average individual loan still remain manageable. In the event that several group members request for RE related loans simultaneously, they could 'graduate' into direct individual borrowers with the bank. Given their repayment history as part of the group, the information asymmetry, hitherto associated with such loans, is eliminated, [36]. Where repossession is technically feasible, the funded asset could serve as collateral, failing which, the loan could still be channeled through the group, thus benefiting from the security offered by group dynamic.

As an additional measure of loan security, the funded asset, typically, the solar home system or the biogas digester, is insured against theft, malfunction, and incidental damage, by the MFI (and reinsured by a reputed insurer). It is also suggested that the borrower's life be insured to secure the lenders' interests in the event of premature death. Sadoulet [35] discusses microfinance repayment insurance to reduce borrower vulnerability and to improve repayment rates, with contractual conditions similar to those laid down by credit card companies, which intrinsically deter moral hazard and adverse selection. It is evident that several collateral substitutes and insurance instruments are available to mitigate the risks associated with the larger loans for RE systems.

7.6 Regulatory Definitions and Market Facilitation

Elimination of institutional rigidities is more significant and appropriate than subsidizing operations [2] and hence, innovative options need to be offered to permit MFIs morph as they grow organically, [41]. Hybrid structures need to evolve from the convergence between formal banking and informal MFI systems, [28]. Regulations need to be adapted to suit growing and deposit-taking MFIs. Organizations originally spun off NGOs and/or registered as trusts, societies or not-for-profit organizations, are neither disposed to, nor are eligible to, undertake commercial borrowing and lending activities. Legislation governing such forms is not appropriate for microfinance, [40]. Non-banking finance companies and chit funds are regulated but it is highly unlikely that a large number of MFIs would graduate to these levels given the higher capitalization and reporting requirements. MFIs are thus caught in a vicious cycle as commercial investors shy away from unregulated and often uncertain business situations. A legal definition, which would fit the space between the stand-alone MFIs on the lower side and (non-banking) finance companies on the higher, would need to be developed. This would entail simplified registration and rating procedures, lower capitalization requirements, low-cost deposit insurance, along with the flexibility to raise equity and debt from multiple sources, and most importantly, permission to mobilize deposits. These larger, stronger and mature MFIs would be in a position to offer larger loan sizes more frequently to their members, as would be necessary for micro infrastructure investments.

7.7   Promoting Eco-Efficiency

Debating environmental conservation might sound exotic to the poor citizens of the developing world, whose primary motivation is that of immediate survival. And yet, unsustainable harvesting of forest produce, aquaculture, cattle grazing, brick making, leather tanning and electroplating, etc., could lead to progressive environmental degradation. MFIs should promote environmental awareness and eco-efficiency among clients and guide them towards adopting environmentally sound technologies, [47]. MFIs, therefore play dual roles, of creating awareness and of financing purchases. RE technologies readily lend themselves to such situations, offering appropriate solutions with limited or no adverse environmental impact.

7.8   Conclusions and Recommendations

"Microfinance is perhaps best thought of as a platform, rather than simply as another intervention. It creates an infrastructure where the poor, previously seen as isolated and without material assets or social capital, can be mobilized in large numbers and provided with finance to participate in economic and social initiatives", [13]. Returns on domestic Renewable Energy applications accrue from cost avoided, as opposed to revenue generated. The larger loans made to assist the purchase of such applications lead to 'quality of life' improvements on the one hand, and to institutional stability and sustainability of the MFI, on the other.

The demand for such loans reflects borrower maturity and her/his liberation from abject poverty and hence could be considered an index of mission compliance for the MFI. Institutional structures forged by bringing together the attributes of formal banking and informal MFI—solidarity group channels would provide for requisite refinance, while keeping costs of intermediation down. MFIs could also serve to educate the potential consumers on the environmental benefits of RE and other technologies. It is worth mentioning that several MFIs and cooperatives in Nepal have lent extensively for the installation of biogas digesters under the

Biogas Support Program[10] (BSP) of Nepal. Similarly, Sarvodaya Economic Enterprise Development Services (SEEDS), a Sri Lankan microcredit institution has actively promoted residential solar home systems for over a decade. Some Indian rural banks have also participated in such lending programs. However, such instances are more the exception than the norm.

An unambiguous definition of the organization's mission, the aims and uses of microfinance [24] when combined with objective social performance measurement, [31] facilitates superior client segmentation, which in turn enables responsive product development, increases efficiency and improves risk management. The role and scope of the MFIs need to be reoriented from *product-centered schemes to client-centered approaches,* [3] custom designing financial services to smaller segments of borrowers. This is especially relevant when banks and mainstream financial institutions seek to expand beyond the urban clusters, and simultaneously, to lower costs of intermediation involved in catering to the hinterland. Banks franchise the matured and relatively healthier MFIs, offering both refinance and fee-based incomes to extend the banks' outreach in the rural areas.

Development Finance institutions and other such 'creatures of special statute' are in threat of outliving their utility, unless they are guided by a mission that itself evolves over time. Such progress needs to be supported and facilitated by appropriate legal and regulatory definitions for MFIs, as they grow into more formal and systematically managed enterprises. Statutory definitions need to include, *or at least not preclude*, the provision of micro loans for RE systems. For Microfinance to evolve to the next stage, to ascertain borrower 'graduation' and to help mainstream RE systems, exogenously imposed restrictions on end-use application of funds, loan size, collateral requirements and the like need to be done away with. Renewable Energy technologies serve to improve the quality of life of MFI clients and thus, need to be integral to the charter of MF programs.

References

1. Asian Development Bank (ADB) http://www.adb.org/Microfinance/default.asp
2. Ahuja R, Jutting J (2004) Are the poor too poor to demand life insurance. J Microfinanc 6(1):1–20
3. Arun T, Hulme D (2003) Balancing supply and demand: the emerging agenda for microfinance institutions. J Microfinanc 5(2):1–5
4. Basu P, Srivastava P (2005) Exploring possibilities: microfinance and rural credit access for the poor in India. Econ Polit Wkly 23:1747–1755
5. Beck T, Demirguc-Kunt A, Martinez P, Maria S (2005) Reaching out: access to and use of banking services across countries. World Bank. http://siteresources.worldbank.org/DEC/Resources/ReachingOutSept9.pdf
6. Biswas S (2003) Housing is a productive asset–housing finance for self-employed women in India. Small Enterp Dev 14(1):49–55

---

[10] www.bspnepal.org.np.

7. Bond P, Rai A (2002) Collateral substitutes for microfinance. Center for International Development, Harvard University
8. Carpio MA (2004) The experience of financial institutions in the delivery of microcredit in the Philippines. J Microfinanc 6(2):113–135
9. Cecelski E (2000) Enabling equitable access to rural electrification: current thinking and major activities in energy, poverty and gender. The World Bank
10. CGAP (2003) Making insurance work for the poor: current practices and lessons learnt, http://www.cgap.org/press/press_coverage23.php
11. CGAP (2004) Case studies in donor good practices, No. 11, http://www.cgap.org/direct/docs/case_studies/cs_11.php
12. Chowdhury IA (2005) Why a year: quote collection. Year-of-Microcredit.org, http://www.yearofmicrocredit.org/pages/whyayear/whyayear_quotecollection.asp#stanleyfischer
13. Counts A (2004) Microfinance and the global development challenge. Econ Perspect 9:1
14. Djamin M, Lubis AY, Alyuswar F, Nieuwenhout FDJ (2002) Social impact of solar home system implementation: the case study of Indonesia (Kolaka, South East Sulawesi). World Renewable Energy Congress VII (WREC)
15. Dunford C (2001) Building better lives: sustainable integration of microfinance and education in child survival, reproductive health, and HIV/AIDS prevention for the poorest enterpreneurs. J Microfinanc 3(2):1–25
16. Easton T (2005) The hidden wealth of the poor: a survey of microfinance. The Economist http://www.economist.com/node/5079324
17. Fischer S (2005) Why a year: quote collection. Year-of-Microcredit.org, http://www.yearofmicrocredit.org/pages/whyayear/whyayear_quotecollection.asp#stanleyfischer
18. Gibbons DS, Meehan JW (1999) The microcredit summit's challenge: working toward institutional financial self-sufficiency while maintaining a commitment to serving the poorest families. J Microfinanc 1(1):131–192
19. Goldberg M, Motta M (2003) Microfinance for housing: the Mexican case. J Microfinanc 5(1):51–76
20. Jansson T (2001) Microfinance: from village to wall street, sustainable development department–best practices series, inter- American development bank. http://www.iadb.org/sds/doc/ MSM113VillagetoWallStreetJANSSON.pdf
21. Jones H, Williams M, Thorat Y, Thorat A (2003) Attitudes of rural branch managers in Madhya Pradesh, India, toward their role as providers of financial services to the poor. J Microfinanc 5(2):139–167
22. Kalpana K (2005) Shifting trajectories in microfinance discourse. Econ Polit Wkly 40(51):5400–5409
23. Kamp J (2003) Small change. Lemniscaat, The Netherlands
24. Lashley JG (2004) Microfinance and poverty alleviation in the Caribbean: a strategic overview. J Microfinanc 6(1):83–94
25. Littlefield E, Morduch J, Hashemi S (2003) Is microfinance an effective strategy to reach the millennium development goals? CGAP Focus Note 24
26. Martinot E (2005) Renewables 2005: global status report. The Worldwatch Institute, Washington, p 30
27. Mayfield JB (1998) Grameen bank replication: lessons learnt. Fall Publishing, Choice Humanitarian, p 2
28. Nair TS (2005) The transforming world of Indian microfinance. Econ Polit Wkly 40(17): 23–29
29. Otero M (1999) Bringing development back into microfinance. J Microfinanc 1(1):8–19
30. Painter J, MkNelly B (1999) Village banking dynamics study: evidence from seven programs. J Microfinanc 1(1):91–116
31. Pawlak K, Matul M (2004) Realizing mission objectives: a promising approach to measuring the social performance of microfinance institutions. J Microfinanc 6(2):1–25
32. Rijsberman F (2004) Understanding the global water crisis. The Economist
33. Rural Microfinance Development Center (RMDC) http://www.rmdcnepal.com

34. Ruben R, Clercx L (2003) Rural finance, poverty alleviation, and sustainable land use: the role of credit for the adoption of agroforestry systems in occidental honduras. J Microfinanc 5(2):77–100
35. Sadoulet L (2002) Incorporating insurance provisions in microcredit contracts: learning from visa®? World institute for development economics research, Discussion paper No.2002/56, United Nations University
36. Satish P (2005) Mainstreaming of Indian Microfinance. Econ Polit Wkly 40(17) April 23– April 29, 2005
37. Saywell D, Fonseca C (2006) Micro-credit for sanitation. Well fact sheet, http://www.lboro. ac.uk/orgs/well/
38. Schreiner M (2001) Seven aspects of loan size. J Microfinanc 3(2):27–47
39. Seibel HD, Torres D (1999) Are grameen replicatons sustainable, and do they reach the poor?: The case of CARD rural bank in the Philippines. J Microfinanc 1(1):117–130
40. Sriram MS (2005) Microfinance and the state: exploring areas and structures of collaboration. Econ Polit Wkly 40(17) April 23–April 29, 2005
41. Sriram MS, Upadhyayula RS (2004) The transformation of the microfinance sector in India: experiences, options and future. J Microfinanc 6(2):89–112
42. Temple FT (2000) Incorporation of renewable energy technology in development projects. World Bank
43. Tucker M, Miles G (2004) Financial perfromance of microfinance institutions: a comparison to performance of regional commercial banks by geographic regions. J Microfinanc 6(1):41–54
44. Ud Dowla A (2004) Microleasing: the grameen bank experience. J Microfinanc 6(2):137–160
45. Varley RCG (1995) Financial services and environmental health: household credit for water and sanitation. http://www.gdrc.org/icm/environ/usaid.html
46. WBCSD (2004) Distributed solar energy in Brazil. Case study, World Business Council for Sustainable Development. www.wbcsd.com
47. Wenner MD, Wright N, Lal A (2004) Environmental protection and microenterprice development in the developing world: a model based on the Latin American experience. J Microfinanc 6(1):95–122
48. Wilkins G (2002) Technology transfer for renewable energy. The Royal Institute of International Affairs, UK
49. Wright GAN, Brand M, Northrip Z, Cohen M, McCord M, Helms B (2002) Looking before you leap: key questions that should precede starting new product development. J Microfinanc 4(1):1–15
50. Wright G (2000) Replication: regressive reproduction or progressive evolution? J Microfinanc 2(2):61–82
51. Yunus M, Jolis A (1998) Banker to the poor. Aurum Press Limited, London

Chapter 8
Epilogue: Rational Exuberance for Renewable Energy

Government does not solve problems: it subsidizes them
Ronald Reagan
40th President of the USA

Ethanol as a fuel/fuel-blend has been a success story in Brazil and select other countries; ethanol is an octane enhancer and fuel-ethanol blends help reduce hydrocarbon, carbon-di-oxide and nitrogen oxide emissions. In addition such fuels reduce emissions of benzene and butadiene, sulphur-dioxide and particulate matter. In summary, ethanol's credentials as a fuel additive with superior environmental outcomes are unchallenged. Sale of ethanol also represents an additional stream of cash-flows for the sugar mills that produce it. However, besides being a fuel additive, ethanol finds application in several other industry sectors, with the potable alcohol segment being among the largest consumers.

Despite the environmental and other advantages of ethanol or other such products, prices are benchmarked against the nearest, practically feasible, substitute by both consumers and producers. Basing on the calorific value of ethanol, oil refining and marketing companies in India have computed that a liter of ethanol should be priced at INR 14 ($\sim$ US 30 cents). In tenders floated in the year 2006, the government invited supplies at INR 21.50 ($\sim$ US 45 cents). In the year 2009, oil companies continued to negotiate with distillers for supplies at INR 25–26 ($\sim$ US 55 cents) per liter of ethanol, beyond which price blending with petrol (gasoline) would prove unviable. In December 2009, a group of federal ministers debating the subject arrived at the conclusion that "ethanol blending was an impractical plan," [6].

The key to this apparently irreconcilable dilemma lies not within the agriculture-sugar-energy sectors but in the potable alcohol industry which makes

---

An extract of this chapter has appeared on Energy Manager, a quarterly publication of the Society of Energy Engineers and Managers, India: July–September 2010 issue.

S. Sunderasan, *Rational Exuberance for Renewable Energy*,
Green Energy and Technology, DOI: 10.1007/978-0-85729-212-4_8,
© Springer-Verlag London Limited 2011

competing claims on the ethanol produced in the country. Manufacturers of potable alcohol procure ethanol from the sugar companies at prices in the range of INR 50–60 ($\sim$ US 125 cents), or over twice the price offered by the government tender. Obviously, ethanol producers see no reason to sell at the low prices offered by the government when the ruling prices in the chemical and potable liquor industries are much higher. Renewable energy operates in the real world, and it cannot be assumed that the conventional theories and incentive structures of economics and business would, for some reason, not apply to renewable energy.

8.1 Positive, Non-Distortionary Policy

Policy objective should be one of achieving environment-friendly energy/electric power generation at least-cost while creating jobs in the process; the technology employed is merely an instrument—a matter of detail. The feed-in-tariff, which is seen to have dramatically grown the German, Italian and a few other markets for renewable energy, and to have created large numbers of 'green jobs' in the process, is not a one-size-fits-all solution—at least not without sufficient customization and long-term certainty thrown in. For instance subsequent to the collapse of the Spanish solar PV market in 2008–09, "the PV sector has been criticized as the most expensive and least efficient of Spain's renewable sectors, with leading figures distancing themselves from the industry," [2]. Generic, misplaced policy and poorly targeted subsidies lead to large-scale wealth destruction and impose unjustifiable burdens on tax payers.

As concluded in Chap. 2, financing mechanisms should be non-distortionary, inexpensive to administer and competitively neutral, enhancing allocative efficiency and not benefiting a few firms or technologies at the expense of others. In contemporary times, *country-specific* policies that seek to stimulate the market but those that quickly run out of steam create *global* cycles of scarcity and glut and do not help retain investor confidence. The rapid expansion of the German market over 2003–2005 drove up solar PV cell and module prices as the supply side had failed to keep pace. Even as production capacities were being augmented, the collapse of the Spanish market in 2008–2009 left accumulated inventories, driving product prices down sharply.

Depending on resource availability, and on the relative attractiveness of various options to harness such a resource, investors should be invited to compete and set up generation and ancillary infrastructure. Effective policy should be *technology agnostic*, i.e., neither favoring nor discriminating against technology options. As discussed in Chap. 5, depending on the stage in the life cycle, grants and subsidies could be provided to support research, development and technology improvement efforts but should not be employed as an instrument of state policy to intervene in specific markets. Likewise, funding from multilateral agencies and grants intended for development assistance could be utilized for skill upgrades, and for other non-commercial purposes. Liberalized entry, fair competition and liberalized exit, encourage efficient utilization of available resources and help firms generate a wealth of service and price options to reach the target markets.

8.2   Fitting the Solution

Policy to encourage alternative fuels and renewable energy sources needs to be framed with a broad perspective, taking into account the relative scarcity of individual elements, the role played by substitutes, and alternative uses for the resource. Constraints facing implementation and sustenance of programs need to be systematically analyzed. Cultural and life style issues need to be incorporated at a very early stage in the policy formulation process. For example, there have been instances of box-type solar cookers being used to store clothing, as the farm workers who had received them left home early in the day and returned after dusk.

The distribution of box-type solar cookers to daily wage earners such as farm labor and construction workers is a classic case of fitting a problem to the solution. Solar PV pumping systems have been supplied to agriculturists at heavily subsidized prices in the Punjab and few other parts of India, so as to convert the perpetual subsidy on power to the farm sector to a one-time occurrence. Because of such deep discounts, a retail private-sector market for solar PV pumping systems has failed to take off. 'Improved' cook-stoves that burn biomass fuels have been designed, developed and demonstrated for decades, but few, if any, of the designs have actually been commercialized and widely disseminated on their own merit.

> Donor agencies and non-governmental agencies involved in the promotion of clean energy solutions should restrict themselves to demonstrating the technology and to creating awareness. The ultimate choice of technology-service packages should be left to the end-users. Even if technology packages are pushed on the strength of subsidies, their continued utilization, and hence the environmental benefits slated to flow from their application, would accrue if, *and only if*, these technology packages are integrated seamlessly into people's lifestyles and work practices.

> Subsidies (capital or interest subsidies), concessions, rebates etc. need be targeted to ensure that they reach the intended beneficiary, i.e., in most cases the user of RE technology. Providing land, low-cost loans, loan guarantees and the like for the set up of manufacturing operations would help swell the manufacturers' margins unless such concessions are passed on as lower product prices. More importantly, subsides should be offered for a finite time window with a pre-defined and publicly announced attenuation pattern.

> In markets with relatively mature banking systems and organized retail networks, policy should envisage employing the existing infrastructure rather than attempting to set up parallel channels for specific purposes such as for the

promotion and distribution of RE technology. Likewise, in markets with weak banking infrastructure, policy should inherently work on strengthening existing channels so as to ensure sustenance beyond the short-term. The creation of special-purpose supply chains leads to bureaucratic entrenchment, and the very channels reciprocally begin to justify the continuance of subsidy and other promotional initiatives, thus stifling the growth of the industry segment and preventing it from maturing to the next stage in the life cycle.

8.3 Policy to Encourage Long-Term Investments

Oregon state's feed-in-tariff contracts for solar PV projects were gone in 900 s. In all, 75 projects received the award in the first round of bidding organized in July 2010—literally within the first 15 min. Solar installers are to receive 65 cents per kWh of electric power delivered to the grid for fifteen years [4], substantially above grid-supplied power priced at about 6–9 cents per kWh, which is largely generated by hydroelectric power stations. In addition, the state has also offered some of the most attractive incentives among American states to encourage the set up of solar PV module manufacturing facilities.

Unfortunately, such generous incentives are not entirely retained by end-users. In the year 2009, the price of the average solar photovoltaic system declined by about 11% while prices of crystalline silicon solar modules declined by 38% [12]. Notwithstanding the collapse of the Spanish market leading to a drop in module prices, and the lower decline in prices of balance-of-systems viz., inverters, regulators, storage equipment and the like, system installers are found to pass on *just enough* of the price decline to consumers to make the installation worthwhile, comparable to the next-best investment opportunity available. The higher margins are cornered by participants in the value chain in the proportion of their relative bargaining power, which is brought about by the relative scarcity of their respective contributions to the value chain. This is also observed in the pricing of wind energy generators, with surpluses from preferential tariffs migrating backwards to turbine manufacturers and site owners, while wind-farm developers (who own and operate the farms and trade the power generated) get to retain just the bare minimum competitive return. Likewise, in the case of the sugar industry discussed in Chap. 3, the surpluses are transferred backwards by government diktat or merely by the farmers threatening to switch crops.

It is also observed that policy makers often underestimate the response to such subsidy programs and rapidly run out of budgetary allocation. In July 2010, Thailand had to halt the intake of solar PV project proposals as the sector had 'overheated' and 500 MW target set for the year 2022 was already surpassed by 500% [7]. Regulatory uncertainty and difficulties with contract enforcement by themselves dissuade investors; policy instability and withdrawal of subsidies subsequent to plant construction could cause irreparable damage to government reputation. For instance, the Spanish wind and solar thermal sectors have had to negotiate a reduction in subsidy so as to eliminate uncertainty of future cash-flows;

it is also believed that the government is likely to restrict the number of subsidy-eligible hours of power generation from solar PV plants [2]. Similarly, the uncertainty associated with the solar feed-in-tariffs in Italy has left investors nervous [9], especially given that power generated by solar PV systems is cost competitive with grid-supplied power in some settings.

In contrast, a shift in bargaining power to the other extreme has been observed in parts of Germany and Texas. It has been reported that, on occasion, wind farms in Germany have generated more electricity than was on demand and consumers *had to be paid* to consume the surplus electric power [11]. In certain situations, utility network managers have ordered that wind turbines be taken offline to help reduce supplies. To start with, subsidies, promotional schemes, rebates and the like are offered to encourage installation and generation, and then consumers are provided with incentives to consume the power so generated!

Arguably, much of this volatility in supply is attributable to the unpredictability of wind flow patterns (or for that matter other resources such as solar insolation for solar PV, or flow in rivers for hydro projects). While the lower prices benefit consumers, such isolation from the grid and the negative prices hurt returns on wind farm investments, in turn discourage further capacity addition.

With the exception of power generated from incineration of biomass, most renewable energy technologies viz., solar photovoltaics, solar thermal systems, small hydro power, wind energy generators operate at low marginal costs. Balancing between the low cost of generation and the intermittency in supply is a major challenge for network managers. The availability of transmission infrastructure to convey the excess power to areas suffering from power deficit would help with smoothening generation and consumption. For instance, wind farm developers in southern India have recorded losses of over US $200 million in the year 2008 alone owing to inadequate carrying capacity of the transmission infrastructure and consequently, wind turbine equipment sales have suffered steep declines [1]. Unsurprisingly therefore, RE technologies in India today constitute over 10% of the installed capacity while contributing just about 4% of the actual power generated [5]. Creating such infrastructure is an integral component of leveling the playing field for competing technologies, of eliminating uncertainty, and of ensuring technology neutrality.

8.4 Attaining 'Grid Parity'

'Grid parity' or being "competitive with conventional power sources" [8], is the often professed, and yet elusive, goal for most RE technologies but is spoken of most frequently in the context of solar PV. In colloquial terms, the phrase stands for the price indifference between grid-supplied power generated from coal, fossil fuels etc. and the power generated by RE technology. On either side of the equation is a price that is determined, administered, regulated, or influenced by policy at the very least. Certain technologies could attain 'grid parity' in certain market settings viz., Japan, Italy etc., purely because the benchmark grid-supply tariffs could be sufficiently high [9]. Additionally, utilities have resorted to the use of increasing-block pricing, under which the marginal price to the household increases as its consumption of electric power rises. Generally, there is no 'cost basis' for differentiating marginal price of electricity consumption and the block-prices are intended to ease the burden on the lower strata of society [3]. Personal wind-energy systems or roof-top solar PV and thermal systems might therefore prove economical, compared to grid supplied power, for certain sections of society: the high-end power consumers, who might then end up benefiting from the subsidies intended to enhance the attractiveness of technology packages.

Additionally, such computations of 'grid-parity' are likely to be fuzzy owing to the intermittency of supplies, the costing of transmission infrastructure, storage systems and the like, not to mention the valuation of environmental externalities. Marginal costs of production cannot meaningfully be compared either, given the vast differences in time-profiles of costs for large power plants and RE technologies. Also, the willingness to pay for the additional unit of electric power (the marginal price) by itself varies by a wide margin: price of power traded in India varied between INR 9 ($\sim$ US 20 cents) in April–May 2010 and INR 2.6 ($\sim$ US 6 cents) in June 2010 [10] and is expected to range between INR 4 and 5 ($\sim$ US 9–12 cents) in the medium-term.

'Grid parity', therefore, is entirely contextual and cannot be the central theme of a promotional program. It is the function of policy to create the right environment to help industry generate multiple options and for the end-users—individual households and businesses alike—to help migrate to a low carbon economy. Smoothly functioning product and financial markets would help allocate investment dollars to the right technologies at the right times. A rigorous analysis of the renewable energy industry and markets, and the consequent design of appropriate incentive structures would help generate the desired rational exuberance for such technologies.

References

1. Allirajan M (2008) Gone with the wind. Businessworld, 11 Feb 2008, p 54
2. Backwell B (2010) Subsidies to be cut for Spain's wind and thermal solar sectors. Recharge News, 5 July 2010. http://www.rechargenews.com/business_area/politics/article219756.ece
3. Borenstein S (2010) The redistributional impact of non-linear electricity pricing, NBER Working Paper No. 15822, National Bureau of Economic Research, March 2010
4. Kanellos M (2010) Oregon's feed-in tariff sells out in 15 minutes. GreentechMedia, 2 July 2010. http://www.greentechmedia.com/articles/read/oregons-feed-in-tariff-sells-out-in-15-minutes/
5. Liang LH (2010) India's renewable energy may triple to 48GW by 2015. RechargeNews.com, 16 July 2010
6. Mahajan AS (2009) Evaporating ethanol plan. Businessworld, 21 Dec 2009, pp 30, 31
7. Recharge News (2010) Thailand halts new solar proposals as sector overheats. 19 July 2010. http://www.rechargenews.com/energy/solar/article221895.ece

8. Singer P (2010) Grid parity: on track to reducing cost/watt. Photovoltaics World 2(2), March–April 2010
9. Stromsta K-E (2010) Italian solar industry left on tenterhooks over new FIT system, Recharge News, 7 July 2010. http://www.rechargenews.com/energy/solar/article220455.ece
10. Subramaniam K (2010) Power: unseasonal crash. Businessworld 21 June 2010, p 19
11. Van Loon J (2010) Windmill boom cuts electricity prices in Europe. Bloomberg 23 April 2010. http://www.bloomberg.com/news/2010-04-22/windmill-boom-curbs-electric-power-prices.html
12. Wicht H, Sheppard G (2010) Survival of the FiT-test. Renewable Energy World, March–April 2010, pp 45–53